INTRODUCTION TO SATELLITE TV

by
Chris Bowick and Tim Kearney

Howard W. Sams & Co., Inc.
4300 WEST 62ND ST. INDIANAPOLIS, INDIANA 46268 USA

Copyright ©1983 by Howard W. Sams & Co., Inc.
Indianapolis, IN 46268

FIRST EDITION
FIRST PRINTING—1983

International Standard Book Number: 0-672-21978-6
Library of Congress Catalog Card Number: 83-60159

Edited by: C.W. Moody
Illustrated by: Jill E. Martin

Printed in the United States of America.

Preface

Technology is advancing at a staggering pace. Forty years ago, the most sophisticated piece of electronics in the home was an AM radio. Today we are presented with video games, home computers, projection television, portable video cassette recorders, and even home satellite TV receivers. It is with this latter segment of consumer electronics that this book is concerned.

The intent of this book is to provide a basic understanding of the home satellite TV system. There will be no in-depth circuit analysis or "how-to" construction segments. Instead, there will be an overall examination of satellite TV in general with particular emphasis on the elements comprising such a system for the home.

Chapter 1 covers the evolution of satellite communications from Arthur C. Clarke's article in 1945 through the successful orbiting of a synchronous communicaitons satellite. There are explanations of modulation, propagation, long-distance communications, and the relaying of information. Satellite system considerations and components are discussed as well as the future of such communications. All of these appear in a lightly technical presentation.

The second chapter begins with business and technical aspects but quickly jumps to a more comprehensive explanation of the satellite television system. The uplink is discussed from the broadcast studio through signal propagation. A generalized view of the inner workings of a satellite is included to show that the simple task of reflecting a signal in space requires a very sophisticated communications platform. The downlink is then the final path of propagation that the satellite signal follows.

Chapter 3 is a discussion of the receiving antennas and low-noise amplifiers required for home satellite TV reception. This includes an explanation of the most widely used antennas. Different types of feeds are presented as well as the mounts that must be used to support and aim the receiving antennas. The anatomy of a low-noise amplifier/converter is presented in some technical detail. Antenna and LNA trade-offs are discussed to complete the presentation and allow the reader to make an educated decision in the selection of these components.

The fourth chapter explains the operation of a satellite TV receiver. The signal path is followed from the downlink through reception, processing, and output. The major functional blocks are discussed to familiarize the reader with this final element in a complete home satellite TV receiving system. Again, no in-depth circuit analysis is attempted.

The Appendix includes areas of interest that will act as reference material. Appendix A is a glossary of satellite TV terminology. Appendix B is an explanation of the Direct Broadcast System (DBS). Finally, Appendix C is an aid to locating satellites and aiming the antenna.

As mentioned earlier, this book is primarily a lightly technical discussion of a highly sophisticated subject. The authors, as electrical engineers whose professional careers were spawned from an interest in home electronics and ham radio, offer you this book as an educational tool and a quiet reference for anyone interested in expanding his or her awareness of satellite TV reception.

CHRIS BOWICK
TIM KEARNEY

To our families. . . .

Contents

CHAPTER 1

Satellite Communications.....................................11
Modulation — Propagation — Long-Distance Communication — Clarke's
Artificial Relay System — Evolution of Satellite Communications — System
Considerations — System Components — The Future of Satellite
Communications

CHAPTER 2

The Satellite TV System 29
A System Overview — The Uplink — The Satellite — The Downlink
—Summary

CHAPTER 3

Receiving Antennas and Low-Noise Amplifiers 59
Why a Dish? — The Anatomy of a Dish — Types of Dish Antennas —
Summary–Antennas — The Antenna Mount — The Feed — Feed Rotators
— The Low-Noise Amplifier — The Converter — The Antenna/LNA
Trade-Off — Summary

CHAPTER 4

The Satellite TV Receiver 105
    Receivers — The Downlink Signal — Channel Selection — The Tuner
    —The IF — Demodulation — The Output Signals — Summary

APPENDIX A

Glossary 121

APPENDIX B

Direct Broadcast System 129
    The DBS Receiver

APPENDIX C

Locating the Satellites and Aiming the Antenna 133

Bibliography 139

Index 140

Chapter 1

Satellite Communications

People have been communicating in one way or another for thousands of years. It is only during the last 100 years, though, that they have begun to develop technologies to greatly improve their communications ability. One of those technologies is electronics, which has been used to develop radio and television as we know them today. In 1945 when television was not yet commercially available, a far-sighted man, Arthur C. Clarke, postulated a satellite communications system. Clarke, a science fiction author, wrote an article which appeared in the October 1945 issue of *Wireless World* entitled "Extra-Terrestrial Relays." In Clarke's words, it "...was received with monumental indifference."

Several years later, at a 1954 meeting of the Institute of Radio Engineers in Princeton, New Jersey, a man named J.R. Pierce spoke not of a dream but of the practicality of unmanned satellites acting as radio-frequency reflectors. Technology wasn't far behind for few will forget Sputnik I, the satellite launched by the U.S.S.R. on October 4, 1957. The United States quickly followed with Explorer I, launched on January 31, 1958. For the past 25 years, it seems that we have sent aloft hundreds of these satellites, each one more spectacular than the previous one.

Let's now go back and examine the concept of radio communications in general. In some ways, radio communications might be likened to an airplane, a train, or a car. The latter are used to transport someone or something over a distance. Each uses a different medium (air, rails, roads) but they all accomplish similar tasks. A radio wave might be viewed as yet another means of transportation, carrying a specific cargo—information.

MODULATION

Information that is to be transported long distances via radio waves must first be added to a radio-frequency signal. This signal is called the *radio-frequency carrier*, or sometimes just the *carrier*. The addition of information to a carrier is called *modulation*, a term familiar to some of you. We have all listened to amplitude modulation (AM) and probably frequency modulation (FM). A television signal contains both AM and FM information. As you might imagine, there are many more types of modulation and in-depth theory involved. It is not our intent to present a technical treatise at this point, but to acknowledge and present the existence of the subject.

Most of you are probably familiar with the noise and static that sometimes accompanies AM radio signals during a storm, or while listening to a weak station in the presence of interference. The reason noise is so noticeable on the AM band is because noise is a function of the amplitude modulation process. It, too, is an amplitude modulated signal. This, of course, makes an AM receiver much more susceptible to noise spikes or "crash" as it is sometimes called. For similar reasons, frequency modulated signals are less prone to the static type noise heard on the AM radio bands. That has a lot to do with the popularity of FM radio for music listening. Another characteristic of FM communications involves something called the *FM threshold*. FM stations are usually audible or they are not. Very rarely will you find a weak or static-prone FM station. Instead, what you will hear is the station continually dropping out. In some communications applications, such signal droppage may not be acceptable. In the case of an Air Force pilot, for instance, it is desirable to maintain some type of communications, even though it may be noisy. As a result, most military aircraft utilize AM radios rather than FM. There are many types of modulation, each with a specific intent, so that many different communications needs can be satisfied.

All types of modulation (AM, FM, etc.) could be used at most any frequency (see Table 1-1). It is because of convention, physical (electrical) requirements, and sometimes law that certain types of modulation appear on certain frequency bands. The information that is to be added to an RF carrier is often added on a percentage basis. In other words, the higher the frequency is, the greater is the space available for the addition of information. FM usually requires more room or band-width than does AM. That contributes to the reasons that the FM broadcast band is higher in frequency than the AM broadcast band. Television stations require even more bandwidth (almost 6 MHz). That is one of the reasons that TV channels occupy the frequency band from 50 MHz to 900 MHz. Microwave frequencies (above 3 GHz and higher) can typically contain tremendous amounts of information, such as a multitude of TV stations or very-high-speed data.

Table 1-1. Frequencies and Modulation Employed

Frequency	Modulation	Use
535–1605 kHz	AM	Broadcast (Radio)
88–108 MHz	FM	Broadcast (Radio)
54–78 MHz		
174–216 MHz	AM & FM	Television
470–890 MHz		

PROPAGATION

The way that a radio wave is affected as it travels in space is referred to as *propagation.* The factors that determine how a wave propagates are its frequency and the medium through which it travels. Depending on their frequency, radio waves can bounce or reflect or bend or be absorbed by foreign objects or even by the medium in which they are traveling.

We should all be familiar with frequency. Radio station WOWO in Fort Wayne, Indiana operates at a frequency of 1190 kilohertz (kHz), or 1,190,000 hertz (Hz). The unit of frequency, hertz (abbreviated Hz), is equal to one cycle per second and was assigned in honor of the German physicist, Heinrich R. Hertz (1857–1894), who discovered the principle of radio wave propagation. Prior to the assignment of hertz as the unit of frequency, station WOWO was said to operate on 1190 kilocycles per second (kc/s). The prefix kilo means thousands, mega means millions, giga means billions, and so on.

The AM band we are familiar with extends from 535 kHz to 1605 kHz. You're probably also familiar with the fact that those signals can fade in and out as a function of the time of day. That fading is because those frequencies of radio signals travel upwards from the surface of the earth to a level of the atmosphere called the *ionosphere*. There, the signals then bounce or reflect back down to earth. The ionosphere is like a shell that completely surrounds the earth and its distance from earth varies depending on its exposure to the sun. This movement of the ionosphere causes the fading we are familiar with in the AM and shortwave bands (see Fig. 1-1).

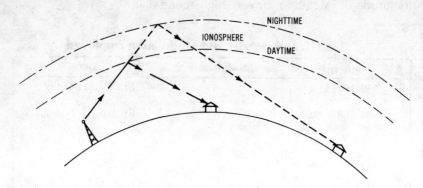

Fig. 1-1. Ionospheric reflection.

This phenomenon may seem to be a detriment; however, shortwave operators have learned to use it to their advantage to allow communications over different distances at different times of day. It does cause some AM stations to adjust their power levels and to redirect their antennas at different times of the day.

Radio waves will propagate via the ionosphere at frequencies up to about 50 MHz. The frequencies above 50 MHz aren't reflected by the ionosphere but instead pass through it. Thus, these frequencies must rely on "line of sight" for communications. These include FM radio, TV stations, microwave radio relays and most satellite systems (see Fig. 1-2).

The frequencies above approximately 1 GHz (gigahertz) are susceptible to another phenomenon called *atmospheric absorption*. That is, as the wave propagates through space, its energy is absorbed in the molecular structure of whatever may be in its path. For this reason, it is difficult to receive these high-frequency signals when there is an obstruction such as a tree in the way. The moisture in a tree tends to absorb the RF energy. At even higher frequencies, near 10 GHz, rain can totally block out communications due to its ability to absorb RF energy.

In summary, remember that the way a radio wave propagates

depends on the frequency and the medium through which it travels. Modulation techniques can vary but usually conform to the type and amount of information being added to the radio-frequency (RF) carrier.

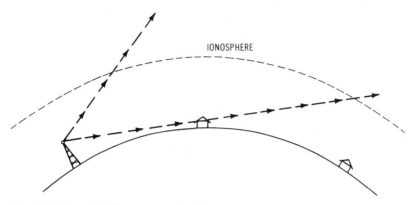

Fig. 1-2. Line-of-sight communication.

LONG-DISTANCE COMMUNICATION

The object of communication as we know it is to transfer information from one place to another efficiently. The engineers who design pocket-size radios must use circuits that will operate from batteries for many hours. In a similar light, the engineers who design a communications system to send a signal from Atlanta to Chicago would be wasting energy if they allowed that signal to also reach New York and Los Angeles. That is to say that for optimum efficiency, the signal should be directed only to the desired locations. It may be a whole section of the country or a very specific region. The medium for transmission can also vary. For example, instead of atmospheric propagation, a wire could be strung from Atlanta to Chicago and direct communications could be established. If a short-wave transmitter were used, large antennas might be used to direct the RF energy (see Fig. 1-3).

In order to send a radio signal over such a distance as Atlanta to Chicago, line-of-sight communications is not possible. It is possible to bounce the wave off of the ionosphere to overcome natural geographic obstacles such as the curvature of the earth. Such propagation could be accomplished by using the ionosphere as a "natural" reflector or by using a series of man-made relay stations. During the Viet-Nam conflict, for example, the U.S. forces discovered that sometimes they could not communicate with a station only 30 miles away because

15

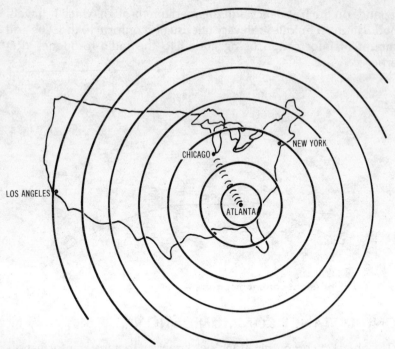

Fig. 1-3. Radio wave propagation.

they were separated by a mountain. An antenna was developed to send the signal almost straight up and bounce it off the ionosphere (see Fig. 1-4). This was necessary because large physical barriers like mountains will not allow radio waves to pass through them. Even in very flat parts of the world, if the separation of two stations is great, even the gentle curvature of the earth can provide a similar obstacle to line-of-sight communications (see Fig. 1-5). Keep in mind, however, that the signals we have been bouncing off of the ionosphere are below 50 MHz.

The "natural" way to send a radio signal, below 50 MHz, from Atlanta to Chicago is to bounce or reflect it off of the ionosphere. Above that, at VHF (very-high frequency), the radio waves are not reflected by the ionosphere. At much higher frequencies, 3 GHz and above, the signals can be easily directed in a narrow beam from one antenna tower to another, 20-30 miles away. This is how much of the telephone company traffic has been handled. Each of these schemes has its purpose but, as in all systems, there are trade-offs. The ionospheric propagation path requires the least sophisticated and least costly equipment. It also transmits much energy to locations where there may be no listeners, and it relies on Mother Nature's natural reflector, the ionosphere, which is not ultimately reliable. The microwave relay system directs its signal very precisely to the intended station. The fact that 30 of these relay stations would have to be built and maintained between Chicago and Atlanta makes this a very expensive system. Remember that technology is not solely responsible for the way we communicate, neither are system cost and revenue. Without a need and a paying customer, very few of these systems would exist.

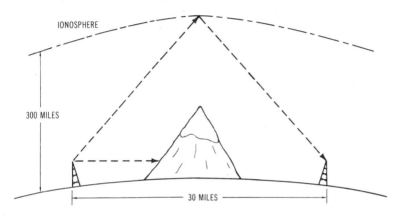

Fig. 1-4. Using the ionosphere to communicate.

What does all of this have to do with satellite communications? The thrust of the previous discussion is that to achieve reliable long-distance communications, both physical and technological barriers must be overcome. Given time and resources, we seem to be improving our technology greatly. But look at the root of the communication obstacles. It seems that to relay radio signals over any great distance, we need to somehow relay or bounce the signals from point to point. The ionosphere provides a near-earth (300 miles) natural reflector for certain frequencies, and microwave relay stations provide an artificial relay system.

CLARKE'S ARTIFICIAL RELAY SYSTEM

Clarke removed himself from the close-in solutions of the previous section. He looked at the ultimate reflector scheme, one limited by geometry and physical size only. His scheme placed a reflector (or communications satellite) so far away from the earth (22,300 miles) that almost half the globe could bounce signals off of one reflector (see Fig. 1-6).

Notice that the areas that cannot use the reflector of Fig. 1-6 are those on the back side of the earth and maybe at a few locations near the poles. However, if two such satellites were placed into orbit, each 22,300 miles above the equator, they could cover all the earth, except those remote polar locations. Well, that is almost true. The people who lived along that longitudinal line right between the two satellites

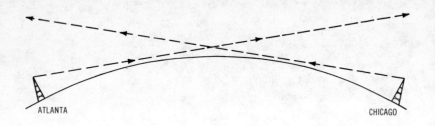

Fig. 1-5. Impossible line-of-sight communications.

would have marginal communications at best. The solution—add a third reflector to the system so we now have three reflector platforms spaced equally apart (120°) along the equator providing global coverage (see Fig. 1-7).

The polar locations would still receive marginal coverage, but because there is little demand for satellite communications there, this is probably a feasible global communications system. Clarke's ingenuity didn't end there. Have any of you asked yourself why a location 22,300 miles above the equator was suggested? It so happens that for a

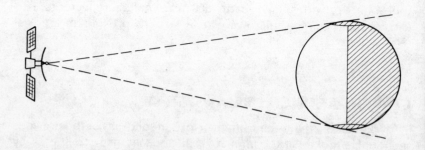

Fig. 1-6. Single satellite coverage.

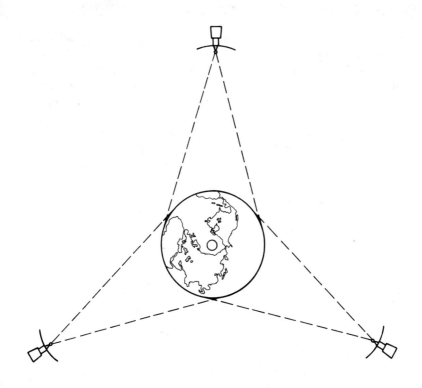

Fig. 1-7. Clarke's proposed three-satellite system.

body to maintain orbit and travel at the same speed as the rotating earth, 22,300 miles is the magic distance. And why on the equator? Well, a body orbiting 22,300 miles above the equator appears to be stationary to an observer on the earth. That is a plus because it means we can utilize an antenna whose position is fixed and always pointed at a satellite. Such a system is much less expensive than a tracking-type antenna system which would be required to follow a moving satellite.

The system proposed by Clarke was a stationary one. He combined what had been learned about communicating with radio waves, orbital mechanics, and far-sighted ingenuity. What he wasn't able to do was fill in the details on how to build the complicated electronics, large antennas, huge rockets, and sophisticated satellites required to implement his system. He did plant the seed, however, and that was sufficient.

EVOLUTION OF SATELLITE COMMUNICATIONS

It was several years after Clarke's article that technology advanced far enough to begin experimenting with artificial means of reflecting a radio wave. Many of the experiments that were tried during those years and judged as failures contributed components and valuable lessons that eventually led to the communications technology we know today.

The U.S.S.R. launched the first orbiting (nonstationary) communications satellite, Sputnik I, in 1957. The U.S. quickly launched its first satellite, Explorer I, in 1958. These first attempts, however, were a far cry from the communications satellites we use today, primarily because many of the problems involved in operating the equatorial orbiting satellite had not yet been solved. Much, however, was learned in these early quests while scientists were attempting to advance certain technologies. For example, while the launch vehicles were being designed and tested, "alternate satellites" were being investigated. The moon is such an alternate. It is a very large passive reflector orbiting the earth, although it is not a stationary reflector. Radio waves were successfully bounced off the moon in 1959–1960 between stations in Holmdel, New Jersey and Goldstone, California. It required very-high-power transmitters, large steerable antennas, and extremely sensitive receivers to complete such a communications link. This type of communication was deemed possible but not really feasible. The costs were high and the system requirements could not be met economically (some not at all). Amateur radio operators (hams) still use the method of bouncing signals off the moon with rather unsophisticated installations compared to those used in 1960. But this method of communication is unreliable at best.

Scientists tried to create their own reflector later in 1960 with the launch of ECHO I, a 100-foot diameter balloon with a metallized skin, orbiting the earth at an altitude of 1000 miles. Some of you may remember lying on the roof of your house at night looking for that shimmering orbiter. This project was an attempt to launch a simple orbiting mirror to reflect radio signals. Another attempt to create an artificial reflector in the sky was project Westford. It consisted of millions of tiny pieces of wire (dipoles) flung into orbit around the earth in an attempt to form a belt-like reflector. Not enough of the signal was ever reflected to deem this a success.

Although these types of communications systems never came into widespread use, they should not be judged as feeble efforts. The intent of placing a passive reflector into space was to greatly reduce the cost of the space-bound element. There were no moving parts, no elaborate launch vehicles, no maintenance, and a relatively indefinite life expectancy (except for ECHO). The object was to place the sophistication in the earth-bound elements (transmitters and receivers) which could be serviced and upgraded with relative ease. Although these early attempts didn't accomplish the communications system that was proposed by Clarke, on February 14, 1963 the first synchronous satellite, SYNCOM I, was put into orbit by the United States.

SYSTEM CONSIDERATIONS

Most of the communications satellites used for television broadcasts maintain a stationary orbit. This allows both the transmitting and receiving earth stations to have fixed antennas which are greatly reduced in cost from those that must follow or track a satellite in nonstationary orbit. Many factors are involved in the selection of an orbital satellite communications system. Keep in mind that for satellite television the stationary orbit is most desirable but that there are many other questions the system designer must answer.

Satellite Positioning

The satellite must both receive signals from the transmitting or uplink station as well as send the downlink signals back to the earth station receivers. Orbit selection has a lot to do with the geographic coverage possible. A satellite orbiting on the equator has the same period of rotation as the earth. But because it is on the equator the areas at both poles have difficulty "seeing" the satellite. Hence, the farther north or south from the equator that an earth station is, the farther it is from the satellite and the weaker the signal is. That is true in general, except for the fact that the satellite can be made to concentrate its transmitted RF energy into specific geographic areas by aiming its large antenna systems. If, for instance, the satellite is only broadcasting to a limited number of receiving stations (or to very localized concentration of earth stations), the antennas can direct signals (called spot beams) to only those stations. However, if there are many receiving stations and they are scattered over a wide area, the satellite must broadcast to these stations over a very wide beam with global or regional coverage. Some of these patterns are illustrated in Fig. 1-8. Similarly, if the satellite receives signals from only one or two uplink stations, then it may direct its receiving antennas to cover only those stations.

It makes sense that a satellite built and financed by the Canadian government would have most of its signal directed to Canadian stations. Notice how the signal strength varies in different areas as shown in Fig. 1-9. The signals are concentrated in certain areas and decrease in level at the fringes of the desired areas. These plots, or maps, are called satellite *footprints.* The patterns are shaped by adjusting the antennas on board the satellite.

Given that a particular footprint has been selected, the satellite must still be positioned so that both the uplink and downlink earth stations can "see" it. A transmitting station in New Zealand could not, for example, relay information to London through a single satellite. The ideal location for a domestic (U.S.A.) satellite coverage is due south from Omaha, Nebraska on the equator. A good location for a satellite meant to relay information from Munich to Chicago is over the Atlantic Ocean.

Thus, it can be seen that even if satellites are positioned on the equator to achieve stationary orbits they still must be appropriately positioned and have their antennas aimed for the desired coverage. As you may have guessed, the equator, also called the *Clarke belt,* is getting crowded with all the satellites located there!

Earth Station Requirements

The size, sophistication, and cost of a satellite receiving installation will vary depending on its intended use. A laboratory or experimental station, for example, will likely have a very large steerable dish antenna (maybe 35 feet in diameter) capable of automatically being aimed at many different geostationary satellites. It may even be capable of tracking polar orbiting satellites. A very large mechanical structure will be required to move the antenna and enable it to function in harsh environments including snow and ice. The electronics will be very sensitive and probably require pressurized cable to carry the signal from the antenna inside to exotic receiving equipment. Little expense will be spared and such an installation may cost more than $100,000.

In contrast, at a private home the antenna will typically be as small as possible, probably six to ten feet in diameter and will have as simple a mounting structure as possible. It, too, should be able to survive in harsh environments, but its electrical performance may be degraded, and there won't be much margin for overall system performance. In other words, the bulk of the equipment sold for these installations will provide minimum performance, hence minimum cost, yet will still be acceptable by the majority of the marketplace. Fig. 1-10 is a typical

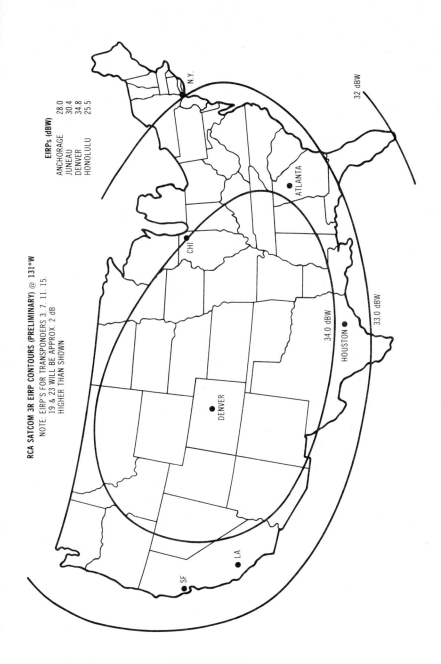

Fig. 1-8. Satellite footprint, SATCOM III-R (in dBW).

23

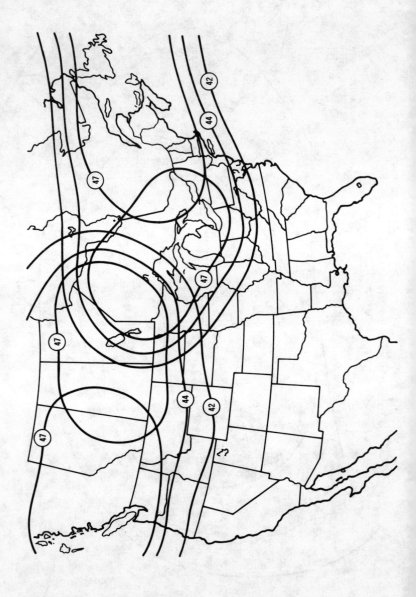

Fig. 1-9. Satellite footprints, ANIK-C (in dBW).

system advertised for use by the consumer. There are many like it available today, but the old caveat still holds true—you get what you pay for!

Satellite Overview

Books could be devoted to the subject of communications satellites. To the casual observer, they simply reflect signals from one place to another. But in reality, they employ the very latest technologies—microelectronics, semiconductor physics, structures, materials, antennas, nuclear energy, etc. In addition, everything on board the satellite must be extremely reliable because no servicing is possible. Everything must operate on low power, be lightweight, and survive being hurled into space. Large antennas and solar energy gathering panels must unfurl when orbit is achieved. Power must be derived for the most part from the sun with batteries storing that energy. Some satellites even use nuclear power sources. All of these intricate systems must operate in an extremely harsh environment—the hard vacuum of space. If something fails, it must be remotely diagnosed and repaired or the system becomes inoperative. It should be evident then why a satellite can cost upwards of 100 million dollars.

SYSTEM COMPONENTS

Let's take a brief look at some of the components that may be required in a satellite communications system. These will be covered in much greater depth later in Chapter 2.

The WARC (World Administrative Radio Conference) has allocated a portion of the radio-frequency spectrum around 6 GHz for the transmission of television signals to a satellite and frequencies around 4 GHz for the transmission of those signals from the satellite to receiving earth stations. The uplink signals have only one destination, a satellite orbiting 22,300 miles above the equator. This means that all that is required is to direct a very narrow beam of RF energy directly to that satellite. This requires a large antenna and a high-power transmitter (see Fig. 1-11).

The antenna does not have to be motorized if it will always be aimed at the same satellite. This can reduce the cost and complexity of an antenna installation. A high-power transmitter is required in order to overcome the loss of RF energy encountered when transmitting a radio signal over such a great distance. In addition, there are certain limitations on the size of the receiving antenna and sensitivity of the receiver

Fig. 1-10. Home earth station.

26

on board the satellite that can be compensated for by having a high-power transmitting station on earth.

The satellite must be capable of receiving signals from several locations on earth and retransmitting them either to a very broad area or a very specific location. The satellite must also be capable of carrying many TV signals, each on its own transponder. Some satellites have as many as 32 transponders. Because these satellites are left unattended in space they must have their own power sources, commonly solar cells and storage batteries. They must also have extremely reliable components capable of operating for seven to ten years.

As the signal from the satellite travels back down to earth, energy is again lost. At present, the receiving station must have an antenna approximately six to ten feet in diameter to collect and direct the signals to a very sensitive amplifier, although advances in technology reduce this size every day. This combination of antenna and amplifier increases the extremely weak signal by as much as 10 billion times (100 dB). A receiver then takes that signal and removes the information contained on the radio-frequency carrier. This information can then be displayed as a television picture.

Fig. 1-11. Transmitting antenna.

Most cable TV operators now offer satellite TV as well as other programming. As the costs decrease, these satellite television-receive-only (TRVO) earth terminals are finding their way into many more places, like churches, schools, apartments, condominiums, hotels, and private homes. By putting the complexity and cost in the satellite and the uplink facility, the cost of the TVRO earth stations is decreasing significantly. Not to mention that as technology advances, the cost of receiving stations will drop even further.

THE FUTURE OF SATELLITE COMMUNICATIONS

Clarke had an interesting idea in 1945—satellites could orbit the earth and reflect signals between stations. It was 12 years before the first communications satellite was launched, and an additional six years before a working relay satellite was placed over the equator in the orbit proposed by Clarke. Soon thereafter, telephone and TV signals were being sent over the ocean instead of through a submerged cable. Then they were used to direct TV broadcasts to broadcasters and cable TV operators all over the country from a satellite located right over (actually due south) the continental United States (CONUS). Electronic tinkerers soon found that they, too, could be listening to and watching all that was on these satellites—movies, sports, and news. It is estimated that there are 50,000 such "listeners" now, and that 30–50,000 new satellite dishes will appear during the next year. Corporations are already proposing systems that would broadcast satellite signals directly into each home. This system is called DBS (Direct Broadcast System) and would reach over two million homes in the U.S. alone. Some speculate that this system is already in place with the number of TVRO stations in operation today. This book will not attempt to look into any crystal balls and predict how things will change in the next six months or in the next six years. Technology has come a long way in the last 25 years since Sputnik and Explorer. It is advancing at a phenomenal rate and still gaining momentum. As a result, it is difficult to predict what lies ahead. Recall that someone born at the turn of the century could have seen the Wright Brothers learn to fly and Neil Armstrong walk on the moon. Someone born in 1950 could have seen scientists learn to launch satellites into orbit and then relay messages across the ocean via synchronous satellites. With that in mind, it is conservative to say that the sky is the limit.

Chapter 2

The Satellite TV System

Now that you have discovered the world of satellite communications and have been introduced to its rather brief history, it's time to turn your attention to the details of just what "black boxes" make up this fascinating satellite TV system. How do stations like HBO, Showtime, Cinemax, The Super Station, CNN, and others make it to your home via satellite and why do they feel that such operation is necessary or even desirable? This chapter will attempt to provide you with a working knowledge of the satellite transmitting and receiving system. We'll even take a look at some typical ground station transmitting equipment to understand its function. The path of the satellite TV signal, from earth to satellite and back again, will be examined in depth. Included is a brief examination of a "typical" satellite and how it operates. Finally, you will be introduced to the satellite earth station, which will begin our in-depth study of the equipment that is most familiar to us all.

A SYSTEM OVERVIEW

As shown in Fig. 2-1, the satellite TV system can be thought of as one giant microwave relay network. The network is made up of a transmitting station, which beams the signal up to the satellite; the signal path to the satellite, which has some important and definable properties of its own; the satellite itself, to relay the information it receives back to earth; the signal path back down to earth, which also has some very

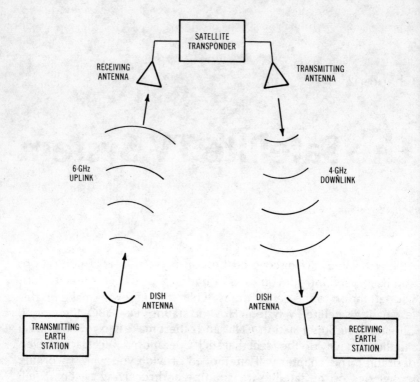

Fig. 2-1. Block diagram of a satellite TV system.

important but definable characteristics; and the satellite receiving system or earth station.

Each component of the system is a vital link in the overall communications process. Quite often, the end users of such a system take it for granted. They just don't seem to realize or comprehend the vast and intricate technologies which comprise the communications link. If something goes wrong within the system, this blissful ignorance can often lead to a never-ending search for a "reliable" service technician when all that was needed was a slight adjustment on the other end of the communications link. It is possible that Mother Nature was the culprit all along and, as a result, the overall poor operation of the system was simply impossible to correct—even for the most skilled technician. Such occurrences are not uncommon when people, who are not familiar with the communications system in use, begin to mistake what could be system problems as problems with their own equipment. Why do you think TV stations, when they are experiencing "technical difficulty," make an extra effort to tell the public not to adjust their TV sets? It's simply because experience has shown that many people assume their sets to be at fault whenever they see a poor quality picture (either that or the station management is trying to prevent viewers from changing the channel). The point being that you need to know as much as you possibly can about the communications system in use, especially if you intend to recognize faults within the system.

Business Aspects

Before we begin to tackle the technical details associated with the satellite TV system, it's important, or at least interesting, to understand some of the business aspects of this communications medium. For instance, who owns the satellites? Who decides the "what, when, and where" for the program material that is carried by a satellite? How much does it cost? Can anyone who has access to the required equipment transmit a signal over the satellite? The answers to these questions lie in the fact that the satellite TV system is in existence to make money for someone! It's as simple as that. The system simply would not exist without the potential for making money.

As was mentioned in Chapter 1, the National Aeronautics and Space Administration (NASA) has launched hundreds of satellites. Many of those satellites were funded and launched by NASA for the sole purpose of space exploration. Other launches, however, were completely funded by other departments of the government and by corporations who were interested in exploring the business potential of space. In these launches, NASA was purely a subcontractor who provided all launch facilities to place the new businesses in orbit. In 1982, for example, NASA provided this service for the SBS and ANIK-C3 satellites, which were carried into space on board the space shuttle Columbia. The satellites themselves are owned by individual corporations. Such is the case with those satellites that are presently in service providing television to millions of homes around the world.

Table 2-1 is a chart which depicts several of the "families" of satellites currently in orbit that carry television programming. The chart indicates the name of each satellite as well as its owner. For example, the WESTAR system is owned by Western Union, SATCOM is owned by RCA, and COMSTAR is owned by Comsat General, a subsidiary of COMSAT, just to name a few. These companies either developed, or contracted with other companies to develop, the satellites they now have in orbit. Once built, the satellites were launched into synchro-

Table 2-1. North American Domestic Satellites

Name	Operator	Launch Date(s)	Number of Transponders per Satellite	Bandwidth (MHz)
Alascom 1 (Satcom V)	Alascom, Inc.	82	24	36
American Satellite Co. 1 to 3	American Satellite Co.	84+	12 & 6 6	36 & 72 72
Anik-A1 to A3	Telesat Canada	72, 73, 75	12	36
Anik-B1	Telesat Canada	78	12 6	36 72
Anik-C1 to C3	Telesat Canada	82+	16	54
Anik-D1 to D2	Telesat Canada	83+	24	36
Comstar D1-D4	Comsat General for AT&T/GTE	76, 76, 78, 81	24	34
Galaxy I to III	Hughes	82+	24	36
GSTAR-1 & 2	GTE Satellite	84+	16	54
Satcom I-II	RCA Americom	75, 76	24	34
Satcom IIIR & IV	RCA Americom	81, 82	24	34
Satcom IR, IIR & "Sixth"	RCA Americom	82-85	24	36
Satmex	Mexico	83	24	36
SBS-1 to 4	Satellite Business Systems	80, 81+	10	43
Spacenet 1-3	SP Communications Co.	83+	12 & 6 6	36 & 72 72
Telstar 3-A to 3D	AT&T	83+	24	34
Westar I-III	Western Union Telegraph	74, 75, & 79	12	36
Westar IV-VI	Western Union Telegraph	82+	24	36
Advanced Westar	Space Comm. Co.	83+	12 6	36 225
USAT-1 & 2	U.S. Satellite Systems, Inc.	85+	10	43

Table 2-1. North American Domestic Satellites (Cont.)

Uplink (GHz)	G/T$_s$ (dBi/K) @ edge	Downlink (GHz)	EIRP (dBW) @ edge	Lifetime (years)	Orbit Location (Longitude)
5.925-6.425	-6	3.7-4.2	34 SSPA	10	143W
5.925-6.425 14-14.5	-4 -2	3.7-4.2 11.7-12.2	34 & 36 42	7.5	
5.925-6.425	-6	3.7-4.2	33 TWTA	7	104, 114, & 114W
5.925-6.425 14.0-14.5	-6 +0	3.7-4.2 11.7-12.2	36 47 TWTAs	7	109W
14.0-14.5	+3	11.7-12.2	48 TWTA	8	112.5, 116, & 109W
5.925-6.425	-6	3.7-4.2	36	8	104 & 109 & 114W
5.925-6.425	-4.5	3.7-4.2	33 TWTA	7	95, 95, 87, & 127.25W
5.925-6.425	-7	3.7-4.2	34.5 TWTA	9	135, 74W +?
14-14.5	+1.6	11.7-12.2	43.1 TWTA	10	103 & 106W
5.925-	-6	3.7-4.2	32 TWTA	8	135 & 119W
6.425	-6	3.7-4.2	32 & 34 TWTA	8	131 & 83W
5.925-6.425	-6	3.7-4.2	34 SSPA	10	139, 66W+
5.925-6.425	-9	3.7-4.2	approx. 35	10	85?
14-14.5	+1.8	11.7-12.2	37 to 43.8 TWTA	7	100, 97, & 94W
5.925-6.426 14-14.5	-5 -5	3.7-4.2 11.7-12.2	34 & 36 SS & TWTA 39 TWTA	7.5	119, 70W
5.925-6.425	-5	3.7-4.2	32 to 34 TWTA	10	87, 95 + 1 more
5.925-6.425	-7.4	3.7-4.2	33 TWTA	7	99, 123.5, & 91W
5.925-6.425	-6	3.7-4.2	34	10	99, 123W
5.925-6.426 14-14.5	-7.4 -5 to +4.4	3.7-4.2 11.7-12.2	33 TWTA 43 to 53	10 10	91W spare @ 79W
14-14.5		11.7-12.2			

Courtesy Satellite Communications Magazine

nous orbit (for a nominal fee) by NASA. As you can well imagine, the expense involved in building and launching a satellite is tremendous. It certainly isn't unusual for a company to spend as much as 100 million dollars to launch a single satellite. Surely, they never would have considered launch if the projected return on investment had not been positive. Investments that are projected to show negative profit (loss) tend to produce angry stockholders.

Once placed in orbit, a satellite can produce revenues for its owner in only one manner. Typically, the owners lease partial use of the satellite to other companies who in turn provide the television services that we are so familiar with. Thus, HBO, Showtime, Cinemax, the Super Station, and others are actually leasing a small portion of the satellite to help distribute their programs. Of course, at the present time, those stations intend their programming only for cable TV operators who pay for those services. The individual earth station owner, however, is presently able to intercept those services free of charge simply because the signals and the equipment to intercept those signals are readily available to almost anyone who wants them. It should be noted that for legal and political reasons, the FCC has refused to rule on the legality of the use of home earth stations. The existence of home earth stations is widespread and a ruling against such installations would be practically impossible.

Often, the owner of a satellite will lease several spaces on the satellite to another company whose sole purpose is to find potential users for that satellite space. Thus, they end up being the "middle man" between the user and the owner. In this case, revenues to pay for the leased channel on the satellite flow from the cable TV subscriber to the cable TV company, from the cable TV company to the satellite user (HBO, etc.), from the user to the middle man, and finally to the owner of the satellite. It's a very involved but very interesting process. Of course, not all monetary transactions within the system flow as just described, but the fact of the matter is that the actual owner of the satellite does get paid for providing his relay station in the sky. And, with the costs to launch the satellite approaching 100 million dollars each, the leased space on the satellite is not inexpensive.

Technical Aspects

Fig. 2-1, shown earlier, presented the overall concept of the satellite television system. In very basic terms, the transmitting station converts video and audio program material to a particular frequency between 5.925 and 6.425 GHz (5925 to 6425 MHz). The signal is then transmitted via a very-high-gain, narrow-beamwidth antenna directly to the satellite, which is to act as the relay station for that signal. The signal received by the satellite is then downconverted to the frequency range of 3.7 to 4.2 GHz and retransmitted back to earth where it is then received at the earth station by a very-high-gain, very-narrow-beamwidth antenna which is pointed directly at that satellite. It should be mentioned here that there are receiving antennas on the market today which do not require that they be pointed directly at the satellite, and we'll be looking at those later in this book. Most antennas, however, do require that they be aimed directly at the satellite of interest. The signal path from the transmitting station to the satellite is generally referred to as the *uplink*. Similarly, the signal path from the satellite to the receiving earth station is called the *downlink*. Both the uplink and the downlink possess some very important characteristics, which will be examined in depth later in this chapter.

Up until this point, we have been referring to the fact that HBO, Showtime, and others lease "space" on a satellite which will carry their programming. This "space" is called a *transponder* and it is nothing more than a single-channel receiver/transmitter (relay station) within the satellite. Different satellites have different transponder capacities. Typically, there are either 12 or 24 transponders on a given satellite (some satellites operating at 12 GHz, also known as the K_u band, have as many as 32 transponders). Thus, some satellites may carry up to 24 different TV signals. The bandwidth of a typical transponder is 40 MHz. This may seem impossible to some simply because all transponders must operate in the frequency range of 3.7 to 4.2 GHz (5.925 to 6.425 GHz uplink), which is a frequency range of only 500 MHz. You could argue that putting 24 transponders in a 500-MHz bandwidth could only allow for about 20.83 MHz per transponder. That argument would be true if it were not for a concept called *frequency reuse*, which allows for adjacent frequency transponders to overlap their bandwidths. This concept will be studied in depth in a later section of this chapter. For now, let it suffice to say that you can, in fact, place 960 MHz of program bandwidth within the 500 MHz allotted.

Thus far in this chapter, a general outline of the satellite communications system has been presented. We've seen that a typical uplink consists of a transmitter, the transmitting antenna, the propagation path to the satellite, the satellite's receiving antenna, and the transponder receiver. Once received by the satellite, the signal is translated to a new frequency and thus begins the downlink. The downlink consists of the satellite transmitter, its transmitting antenna, the signal's propagation path to earth, the earth station receiving antenna, and all associated receiving equipment. The remainder of this chapter will be devoted to examining all elements of the uplink and the downlink in detail except for the satellite earth station itself, which is the subject of Chapters 3 and 4.

THE UPLINK

Ever since television was invented, TV stations have been pushing for larger and larger viewing audiences, much as a magazine publisher constantly tries to increase his magazine's circulation. The larger the viewing audience, the more the station can charge for commercials. There is even a chance that a few regional or national advertising accounts might be won if the station's following is large enough. In years past, the only real way for a station to increase its coverage area was to increase its output power, increase its antenna height, or to use a terrestrial microwave relay system as shown in Fig. 2-2. Typically, a large

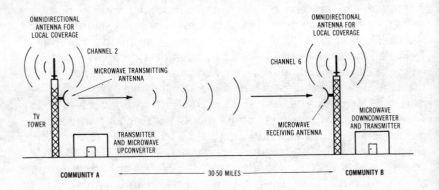

Fig. 2-2. Terrestrial microwave relay system.

increase in transmitter power or antenna height, or both, could only net a small increase in coverage area. That fact, combined with the fact that the typical TV station's coverage area is usually the most densely populated area in the immediate vicinity anyway, means that a small increase in the coverage area for a station does not necessarily add very many people to the viewing audience. Thus, the law of diminishing returns becomes manifest very quickly. What the TV stations needed was a way to reach other very densely populated areas in a very cost-effective manner.

For a while the use of the microwave relay network (Fig. 2-2) was the answer. In this system, the TV signal was transmitted in the normal fashion for its primary coverage area while simultaneously being upconverted to a microwave channel and being beamed to a second densely populated area. At the other end of the microwave link, the signal was downconverted to a normal TV channel and transmitted as a standard TV signal for reception in the home. This approach, though certainly feasible, does have its drawbacks. For instance, if multiple coverage areas were desired (as shown in Fig. 2-3), the system would require multiple microwave transmitters and receivers. Not only would such a system be very expensive for extensive areas of coverage, but repeated upconversion and downconversion of the TV signal can cause significant signal degradation which cannot be corrected. Such signal degradation would show up as noise in the TV picture. In addition to the additive noise problem and the requirement for multiple expensive microwave relays, there is also the problem of having to provide high-power VHF TV transmitters and antenna systems at every distribution point. As you can see, for large distribution systems the cost really begins to mount.

Consider, on the other hand, a system that could cover practically the entire contiguous Unites States using only two transmitters! One of these transmitters is located at the originating facility, while the other is

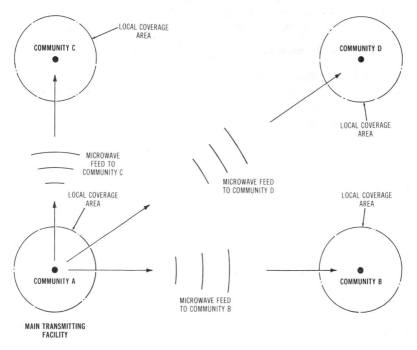

Fig. 2-3. Multiple microwave relay system.

located on a satellite in geosynchronous orbit approximately 22,300 miles above the earth. Such is the system that is the subject of the following paragraphs.

The Uplink Broadcast Facility

The purpose of any TV broadcast facility, whether it is a facility that transmits "over the satellite" or one that is simply your local TV station, is to take video and audio program material and to convert that information into a form that can be transmitted long distances with the ultimate goal of displaying that information on a remote TV. Historically, with local television facilities, that information was converted to radio frequencies in the VHF range (channels 2–13) or the UHF range (channels 14–83) using a modulation process called *vestigial-sideband* (*VSB*) transmission (a form of amplitude modulation) for the video information and frequency modulation (FM) for the audio information. Standard TV sets are presently equipped to receive such transmissions. However, they are not equipped to receive transmissions directly from the satellite for several reasons. This will become evident as we begin to compare the two systems.

The VHF/UHF Transmitter

Fig. 2-4 is a simplified block diagram of a typical VHF or UHF TV transmitter. Note that this block diagram could be adapted to explain the operation of most any of the local TV stations in the country. As shown in the diagram, the baseband video information is first processed through several functional blocks of circuitry whose primary purpose is to precondition the signal to compensate for irregularities in the intended transmission path, including the television set. Once the video information has been processed, the signal is then fed to a mixer or video modulator which upconverts the video signal to an intermediate frequency of 45.75 MHz. The signal at this point is an amplitude-modulated double-sideband (DSB) signal, which is then fed to a device called a *vestigial-sideband filter* to produce the required vestigial-sideband signal. Fig. 2-5 is a pictorial of the differences between an amplitude-modulated double-sideband signal and a vestigial-sideband TV signal. Note that for a vestigial-sideband signal, a portion of the carrier and practically all of one sideband are eliminated. Thus, the bandwidth required for each TV signal is reduced to almost half that which would be required for a DSB signal. This fact allows for twice as many television signals in a given frequency range, which is really the only reason that vestigial-sideband transmission is used in place of DSB.

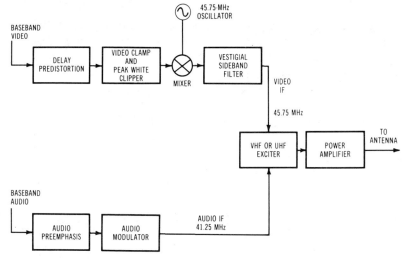

Fig. 2-4. Block diagram of a typical VHF or UHF TV transmitter.

Now let's turn our attention to the path that the audio signal takes to get to this same point in the block diagram (Fig. 2-4). Baseband audio information is first amplified and placed through a preemphasis network which is used to increase the amplitude of the high-frequency components of the audio signal. This technique is used in FM broadcasting to increase the signal-to-noise ratio of the received signal. At the receiver, the signal will then be placed through a deemphasis network to reduce those frequency components to their previous values. It isn't necessary that you understand how this process works, only that it exists. Once preemphasis is applied to the audio signal, the

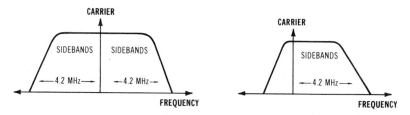

(A) Double-sideband amplitude modulation.

(B) Vestigial-sideband amplitude modulation.

Fig. 2-5. Spectrum of AM DSB versus VSB signals.

39

signal is then applied to the audio modulator. The output of the audio modulator is a frequency-modulated signal with a carrier frequency of 41.25 MHz.

The video and the audio IF signals are combined and then applied to a VHF or UHF exciter, which converts the signals to the assigned output channel frequency and also amplifies those signals to a level suitable for driving a power amplifier to its required output level (in some transmitter schemes, the video and audio paths remain completely separated all the way to the antenna). This output level may be anything from 100 watts to 5,000,000 watts or more depending upon the station's needs as determined by the FCC. The signal we wind up with is a vestigial-sideband video signal and a frequency-modulated audio signal in the frequency range from 54 to 216 MHz for VHF channels, or from 470 to 890 MHz for UHF channels. Output channel frequency assignments for all VHF and UHF channels are shown in Table 2-2. Both the video and audio signals have had some special signal processing provided prior to transmission, such as clamping and white-clipping of the video signal and preemphasis of the audio signal. These points as well as the frequency assignments and types of modulation are important to remember as we begin to compare this system with that of a typical satellite TV broadcast station.

Satellite TV Broadcast Stations

Fig. 2-6 is a simplified block diagram of a typical satellite TV broadcast facility. As we go through the explanation of each block's function, pay particular attention to the differences between this system and that of the VHF and UHF broadcast facility. The ultimate goal here is to provide you with the knowledge of how the system works as well as the understanding of why satellite TV signals are different from those that you might already be familiar with.

Television signals distributed "over-the-satellite" are quite unlike those of your local TV station. One of the most obvious differences that we have already discussed, of course, is the frequency of the transmitted and received signal. Fig. 2-7 shows the frequency plan of the satellite TV system for the C band. (C band is a shorthand reference to the band of frequencies from 3.7 to 6.425 GHz, some of which is presently being utilized for TV transmission.) As is shown in the diagram, the originating station's transmitter must operate on a particular channel in the frequency range from 5.925 to 6.425 GHz. This fact alone prevents a standard television from receiving these transmissions. But there are even more significant differences between the two systems.

Table 2-2. VHF and UHF Channel Assignments

Freq (MHz)	Channel	Freq (MHz)	Channel
LOW-BAND VHF		**UHF BAND (Cont.)**	
54–60	2	638–644	42
60–66	3	644–650	43
66–72	4	650–656	44
76–82	5	656–662	45
82–88	6	662–668	46
HIGH-BAND VHF		668–674	47
174–180	7	674–680	48
180–186	8	680–686	49
186–192	9	686–692	50
192–198	10	692–698	51
198–204	11	698–704	52
204–210	12	704–710	53
210–216	13	710–716	54
UHF BAND		716–722	55
470–476	14	722–728	56
476–482	15	728–734	57
482–488	16	734–740	58
488–494	17	740–746	59
494–500	18	746–752	60
500–506	19	752–758	61
506–512	20	758–764	62
512–518	21	764–770	63
518–524	22	770–776	64
524–530	23	776–782	65
530–536	24	782–788	66
536–542	25	788–794	67
542–548	26	794–800	68
548–554	27	800–806	69
554–560	28	806–812	70
560–566	29	812–818	71
566–572	30	818–824	72
572–578	31	824–830	73
578–584	32	830–836	74
584–590	33	836–842	75
590–596	34	842–848	76
596–602	35	848–854	77
602–608	36	854–860	78
608–614	37	860–866	79
614–620	38	866–872	80
620–626	39	872–878	81
626–632	40	878–884	82
632–638	41	884–890	83

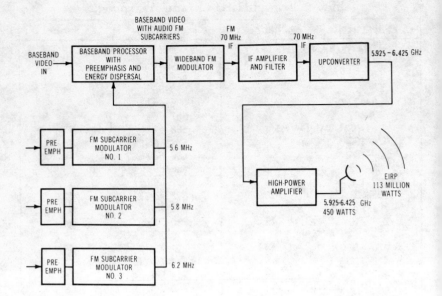

Fig. 2-6. Block diagram of a satellite TV broadcast transmitter.

Referring again to the block diagram of Fig. 2-6, we see that the baseband video program material is applied to an FM modulator as opposed to the AM modulator and vestigial-sideband filter of the terrestrial television transmitter. The output of this modulator is a wideband FM signal quite unlike the vestigial-sideband signal previously discussed. Prior to frequency modulation, however, the video waveform is first processed through a preemphasis network which boosts the signal level of the high-frequency components of the video waveform. This is done in much the same manner, and for the same reasons, that the audio waveform of the standard television signal was preemphasized. That is, preemphasis of an FM signal prior to transmission improves the overall system's signal-to-noise ratio by compensating for certain noise characteristics of the FM demodulator in the receiver. More specifically, as its frequency of operation increases, an FM demodulator becomes noisier. That is, it tends to add more and more noise to the high-frequency components of its demodulated output and thus lowers the signal-to-noise ratio for those components. This characteristic is simply compensated for in the transmitter by increasing the signal level of the higher-frequency components of the baseband signal. The preemphasis is removed in the earth station receiver, after the FM demodulator, by a deemphasis network.

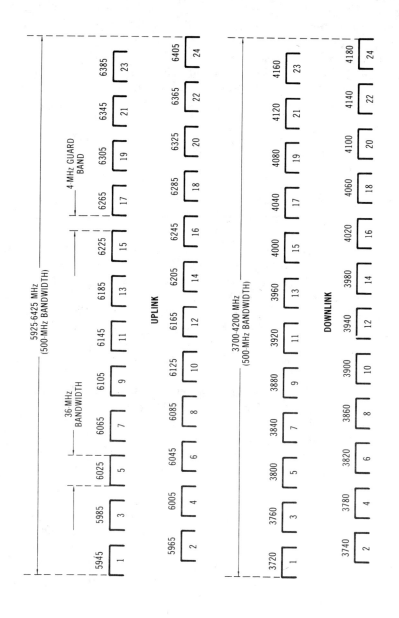

Fig. 2-7. Uplink and downlink frequency plans.

43

In addition to preemphasis of the video signal, certain other signal processing is required prior to modulation. For instance, the so-called "energy dispersal" waveform must be added to the baseband video signal. The dispersal waveform is a triangular waveform running at a frequency equal to the video frame rate, 30 Hz. This signal is synchronous with the video frame rate, and is used to broaden the spectrum of the frequency-modulated RF signal. Without the energy dispersal waveform, and with low average picture levels, the primary component of the transmitted FM signal would be the horizontal synchronization pulses and the 3.58-MHz color subcarrier, which are both present in the composite video signal. This concentration of energy in the transmitted RF signal tends to cause problems with interference to terrestrial microwave links such as those used by the Bell system. To eliminate the concentration of energy, and thus eliminate the associated problems of interference, the dispersal waveform is added to ensure that the spectrum of the transmitted signal is spread out over at least some (about 2 MHz) of the allowed transponder or channel bandwidth. Another feature of the energy dispersal waveform is that it helps to prevent interference among the many TV signals which are carried by a satellite on its various transponders. Of course, at the earth station receiver, this dispersal waveform must be removed from the baseband video signal or it will show up as a 30-Hz flicker on the TV screen. The removal is accomplished by a device called a *video clamp*, which we will study in Chapter 4.

The audio signal, as shown in Fig. 2-6, is also an FM signal. Thus, prior to modulation it too must be applied to a preemphasis network. Once preemphasized, the audio signal is applied to the audio FM modulator which converts the baseband audio signal to an FM signal centered at any one of several carrier frequencies. These frequencies are referred to as *audio subcarriers,* and are typically centered at a frequency between 5 and 8 MHz. Some standard frequencies for audio subcarriers presently in use are 5.6, 5.8, 6.0, 6.2, and 6.4 MHz.

Many earth station transmitters are capable of accepting several independent audio signals. This is accomplished by applying each audio signal to its own subcarrier modulator. The output of each modulator is, of course, at a different subcarrier frequency. All audio subcarriers are then summed with the processed baseband video signal and applied to a wideband FM modulator. The output of the FM modulator is an FM signal centered at 70 MHz with a modulation bandwidth of approximately 36 MHz.

The 70-MHz IF signal is then applied to an upconverter which, after several conversion processes, produces a signal in the 5.925 to 6.425-

Fig. 2-8. Typical transmitting antenna.

GHz range at a level which is capable of driving a klystron high-power amplifier (HPA) to full power output. Though the output of the HPA varies from station to station, a typical output level would be in the vicinity of 450 watts. This is the approximate signal level applied to the transmitting antenna.

The typical transmitting antenna, as shown in Fig. 2-8, isn't much different from the typical earth station receiving antenna. It, too, is a very-high-gain, narrow-beam dish antenna that is aimed directly at the satellite that is to be utilized for relay purposes. The gain of such an antenna can typically be as high as 54 dB, which means that the power actually radiated from the antenna effectively increases by a factor of 54 dB. For an input level to the antenna of 450 watts, this means that the effective radiated power from the antenna approaches 113 million watts! This may seem like an incredible amount of power to blast at a satellite, but as you will soon see, every bit of this power is needed if you wish to provide a usable signal level to the satellite. Every antenna can be characterized by its effective radiated power, sometimes called *effective isotropic radiated power,* or EIRP. This quantity can be calculated for any transmitting antenna by multiplying the input power to the antenna by the gain of the antenna expressed as an absolute number (as opposed to a gain expressed in dB).

There are, of course, many other accessories available for the earth station transmitter which will not be covered here. These devices monitor the status of the transmitting equipment to make sure that the transmitting facility is functioning properly at all times. Transponder time is at a premium, and any loss of signal through the transponder due to equipment malfunction at the transmitter simply cannot be tolerated. For this reason, it is not unusual to see completely redundant (backup) transmitting facilities that will automatically replace the primary facility in the event of equipment malfunction.

As you can see, a television signal transmitted from an earth station transmitter does not even remotely resemble that of your local TV station. Therefore, to receive these signals, you must use an earth station receiver capable of converting these signals to a format that your standard TV set can use. Such a receiver is the subject of Chapter 4.

The Uplink Propagation Path

In the last section, it was mentioned that it certainly would not be unusual to find an earth station transmitter with an EIRP of greater than 113 million watts. But why should we ever need that much radiated power? The reason is simply this: Due to several factors, there is a tremendous amount of signal lost in the propagation path between the earth station transmitting antenna and the receiving antenna located on the satellite. These losses, combined with the possibility of further loss due to the performance of the satellite's receiving antenna and the orientation of its antenna with respect to the earth station transmitting antenna, make for a very weak signal at the input of the satellite's transponder.

As the 6-GHz television signal makes its way from the transmitting antenna to the satellite, several factors combine to decrease the effective power of the signal. One such factor is simply called *atmospheric attenuation*, which, quite frankly, is practically nonexistent at 6 GHz. This type of attenuation occurs through absorption of the RF energy into the molecular structure of the atmosphere and is normally much less than 1 dB. Atmospheric attenuation does increase with heavy rain but rarely exceeds the 1-dB mark. Though atmospheric attenuation contributes only a very small percentage to the overall loss of the propagation path, it is important to recognize because the same contribution occurs in the downlink propagation path and thus effectively doubles the atmospheric loss for the entire link. For an earth station receiver located in a marginal signal level location, atmospheric attenuation could mean the difference between a good picture, or no picture at all on the TV screen.

The primary contribution to the loss of signal between the transmitting antenna and the satellite's receiving antenna is called *space loss,* or *spreading loss,* and will typically amount to approximately 199 dB of attenuation to the transmitted signal. Space loss can be best described as loss which occurs due to the spreading of the spherical wavefront of the RF signal as it propagates from transmitting antenna to receiving antenna. This phenomenon is pictured in Fig. 2-9. Note that the wavefront, as it radiates from the point source (antenna) shown in the diagram, spreads out over a wider and wider area. Thus, the concentration of RF energy is not as great as it was when it left the point of origin. The farther the signal has to travel, the more its wavefront spreads, and the less dense is the concentration of RF energy in the signal. Though all atmospheric propagation paths exhibit space loss, it is particularly important in geosynchronous satellite communications due to the enormous distances covered by the RF signal (approximately 22,300 miles one way). Space loss is proportional to the distance squared, which means that if you double the distance to be covered by the RF signal, all other things being equal, the space loss to that signal increases by a factor of four. This type of loss adds up in a hurry and can approach 200 dB on certain uplinks.

Another source of loss in the propagation path to the satellite relates to the orientation of the satellite's antenna with respect to the

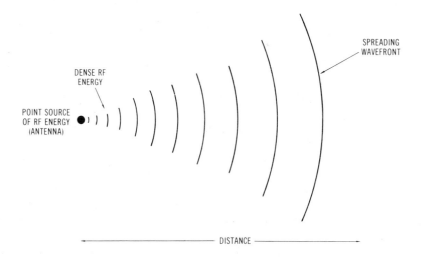

Fig. 2-9. Space loss due to spreading of the spherical wavefront.

transmitter's antenna. This is called the *polarization efficiency* of the propagation path. The signal transmitted from an earth station transmitter to a satellite is either *horizontally* or *vertically* polarized, meaning that its electric field is oriented in either the horizontal or vertical plane, respectively. For optimum efficiency, any receiving antenna that is to receive a horizontally or a vertically polarized signal should itself be oriented in the same plane as the signal that it is to receive. In other words, for best results a horizontally polarized signal should be received by a horizontal antenna. Likewise, a vertically polarized signal should only be received by a vertical antenna. Theoretically, at least, it would be impossible to receive a signal polarized in one direction with an antenna that is oriented perfectly orthogonal (in the opposite polarization) to that signal. Thus, theoretically, a horizontal antenna would provide infinite attenuation to a vertically polarized signal. Infinite attenuation, of course, does not occur, but for orthogonal polarizations it is possible to achieve 20 to 30 dB of attenuation. It is easy to see, then, that when receiving a vertically polarized signal, any misalignment of the vertically polarized antenna on the satellite will cause some loss to the received signal. We would certainly like for that type of loss to approach 0 dB, but in reality, it does have some finite value, typically much less than 1 dB. Again, this is not a significant amount, but under poor signal conditions, it could definitely cause problems.

THE SATELLITE

As was discussed in Chapter 1, in 1945, Arthur Clarke wrote that a satellite orbiting the correct distance from earth would make one revolution every 24 hours, meaning that it would rotate with the earth and would appear to remain in a fixed position with respect to a particular geographic location on earth. He further speculated that three such satellites spaced 120° apart could provide television and other forms of communication to the entire planet. This type of communication is today a reality. Most individuals do not realize how often they have been involved with a satellite processed signal, either actively through an over-the-satellite telephone conversation, or passively at the end of a cable TV system. Much of each network television newscast is either live or taped via satellite. Seldom, when we hear the words "live via satellite," do we even stop to think about it anymore. It has become second nature to us all. It's hard to believe that the signal we are viewing on our television sets has traveled well over 45,000 miles before we ever get to see it!

The following paragraphs will attempt to explain the operation of a typical geosynchronous communications satellite used to carry television signals. As in earlier sections, we will remain with the block diagram approach and will not attempt to describe the actual circuitry involved.

The Overview

As stated earlier, the purpose of any communications satellite is to act as a geosynchronously fixed orbital relay station in order to provide reliable communications over very long distances. For television signals, this satellite is nothing more than an improved version of the terrestrial microwave relay system shown in Fig. 2-2. That is, each system distributes a signal from point A to point B; however, "point B" for the satellite system may include much of the contiguous United States.

The satellite itself is a conglomeration of electronics, rocketry, microwave circuitry, plumbing, solar-energy gadgetry, and computer wizardry that is expected to have a lifetime of approximately 8 to 10 years in the rigors of space. Not only must the satellite house the necessary electronics to perform its assigned task of relaying television program material to your home, but it must also contain plenty of housekeeping material to help sustain its life for such a long period of time. For example, much of the electronics on board the spacecraft has nothing to do with television. Instead, this housekeeping equipment monitors the day-to-day operation of the spacecraft and reports its findings back to a control facility on earth. Such monitoring and reporting is called *telemetry*. The facility that collects this information is usually owned and operated by the same organization that owns the satellite. Once the ground station has received an indication, through telemetry, that an abnormal condition exists on the spacecraft, quite often it is possible to send a command to the satellite to correct the situation. The satellite must, for example, carry 8 to 10 year's worth of fuel on board to make minor corrections to the spacecraft's attitude and orbit in order that long-term drift problems can be counteracted. The commands to correct the position of the satellite are initiated in the ground station tracking facility and transmitted to the spacecraft once its position is found to be out of tolerance. Of course, without rocket thrusters neatly positioned around the spacecraft, such attitude control would not be possible. Thus, once in orbit, the need for rocketry does not diminish.

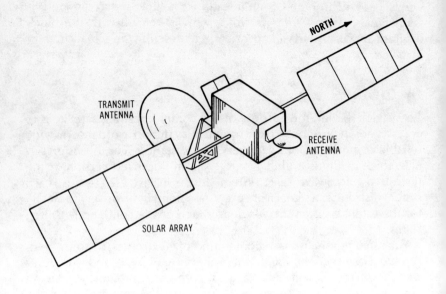

Fig. 2-10. View of satellite with solar panels deployed.

Solar panels are another major component of the satellite. Once the spacecraft reaches its geosynchronous orbit, one of the first orders of business is to deploy this array of panels designed to collect the sun's rays and to convert that radiant energy into a usable source of electricity for all of the electronics on board the craft (see Fig. 2-10). The solar panels typically dwarf the actual electronics on board the satellite due to the immense surface area required to collect sufficient energy to power the satellite's transmitter. Power simply could not be supplied by batteries alone, especially with the need for a minimum service life of eight years.

In summary, a satellite is packed with electronics and perishables, not all of which directly relate to the transmission of television signals. Instead, many of these perform housekeeping chores that increase the satellite's service life well beyond that which would have otherwise been possible. So, now that we have discussed housekeeping equipment, we will turn our attention to the electronics housed in the satellite which is specifically tasked with relaying television program material and other video/audio services from the earth station broadcast transmitter to the earth station receiver.

The Satellite Transponder

Fig. 2-11 is a simplified block diagram of a satellite transponder. Though as many as 24 such transponders may exist on a given satellite, a block diagram depicting all of them would be both redundant and difficult to follow. For this reason, only one will be shown. The other 23 transponders operate identically to the one that is to be described in the following paragraphs, except that their frequency of operation and their polarization vary.

As previously described, the purpose of the satellite transponder is to receive uplink transmissions from an earth station broadcast facility, and to convert that signal to a new frequency prior to transmitting it back to earth. The conversion from the 6-GHz uplink band to the

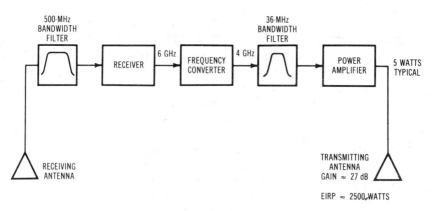

Fig. 2-11. Block diagram of a satellite transponder.

4-GHz downlink band is accomplished according to the frequency plan shown in Fig. 2-7. Note that each transponder is allowed a channel bandwidth of 40 MHz. Of that allotted bandwidth, 36 MHz is used for actual program material, while the additional 4 MHz is used as a *guard band* to help alleviate any potential interference between transponders.

While the frequency plan for the 12-transponder satellite is fairly straightforward and indicates definite frequency assignments for each transponder's transmit and receive functions, such is not the case for the 24-transponder satellites. In this case, 12 additional transponders have been added, virtually on top of the original 12, without decreasing the bandwidth of each transponder or increasing the bandwidth of the entire system. In fact, if you look very carefully at the frequency plan shown, you will notice that the odd and even numbered transponder frequency assignments overlap one another. Such a frequency plan is possible because many of today's satellites make use of orthogonal polarizations for adjacent channel transponders.

Remember from our previous discussions that the signal incident upon a satellite's receiving antenna may be either horizontally or vertically polarized. If the polarization of the receiving antenna is orthogonal to that of the incident wave, then the polarization efficiency of the receiving antenna will theoretically be zero, and no signal power will be received by the antenna. Therefore, if signals on adjacent transponders have orthogonal polarizations, even though their bandwidths overlap, they can effectively be isolated from each other simply by providing each transponder with an appropriately polarized receiving antenna. This frequency plan is called *frequency reuse*, and it effectively doubles the available number of transponders in a given bandwidth. Don't let the technical details bog you down at this point. The important thing to remember is that the polarization of a signal with respect to its receiving antenna has a very pronounced effect on the amount of signal received by that antenna. For optimum operation, the polarization of the receiving antenna should exactly match that of the received signal. Even the slightest mismatch can have a very pronounced effect on system operation. This effect is true on both the uplink and downlink signal paths.

As shown in Fig. 2-7, for a 24-transponder satellite, transponder No. 1 has an uplink (receive) center frequency of 5.945 GHz and a downlink (transmit) frequency of 3.720 GHz. Transponder No. 2 for such a satellite is located only 20 MHz away. Note, however, that for a 12-transponder satellite, transponder No. 1 is located at the same frequency as before, but No. 2 is located 40 MHz away. Such are the frequency plans for a typical satellite which carries television transmissions. As you can see, frequency reuse certainly makes good use of the available RF spectrum.

Referring again to the simplified block diagram of Fig. 2-11, we see that the transponder is nothing more than an elaborate frequency converter. The only thing that really makes it special is that it happens to be located on a satellite approximately 22,300 miles above the equator. All 6-GHz signals incident upon the satellite's receiving antenna are first processed through a broadband 500-MHz filter covering the entire uplink bandwidth. This filter, though broadband, does eliminate unwanted microwave energy (outside of the uplink's bandwidth) from each transponder's input. The signals are then applied to each transponder's receiver circuitry. This block consists of a low-noise, high-dynamic-range amplifier (similar to the LNA of Chapter 3)

52

which is used to amplify the very weak signals present at the satellite's receiving antenna. A substantial amount of filtering is provided at this point to eliminate or greatly reduce the possibility of adjacent channel interference which could cause significant signal degradation. Since the possibility exists that two unrelated signals with adjacent, or near adjacent, transponder assignments may be present at the satellite antenna input, each transponder must provide sufficient selectivity to eliminate the undesired signal. This selectivity is provided by filters in the transponder's receiver in addition to the possible selectivity which may be provided by orthogonal polarizations on adjacent transponders.

The filtered 6-GHz signal is then applied to a frequency converter (mixer) which translates the signal to the 4-GHz range for transmission back to earth. Once translated to its new frequency, the signal is filtered to eliminate any spurious components which may have been added to the signal due to the nonlinear mixing process of the frequency translator, and is then applied to a microwave power amplifier. The power amplifier is similar to, and provides the same function as, the transmitting earth station's power amplifier; that is, it boosts the signal to a level suitable for transmission. Due to the satellite's limited source of electrical energy, which results from the limited surface area of the satellite's solar panels, the output power applied to the satellite's antenna is usually on the order of 5 watts. Such an output power is substantially below that which we have already seen for a typical ground station. Thus, on its return path to earth, the signal has already started off much weaker than its uplink counterpart. To further compound the problem, the satellite's transmitting antenna has much less gain than the earth station transmitting antenna. This is true because the satellite must be capable of transmitting a signal that will cover a very large geographical area. In order for the signal to cover such a large surface area, the antenna must have a very wide beamwidth and, therefore, it cannot focus its radiated energy into an intense beam as did the earth station transmitting antenna. Typically, the gain of the satellite transmitting antenna will approach 27 dB. This means that with an input power to the antenna of 5 watts, the EIRP from the antenna will be approximately 2500 watts, well below the 113-million-watt EIRP of the uplink path.

The geographical area covered by the satellite's signal is called a *footprint* on the surface of the earth. A sketch of such a footprint is shown in Fig. 2-12, and a sample footprint for SATCOM III-R is shown in Fig. 2-13. Note that the contours of the footprint indicate approximate signal levels at various points on the earth's surface. Such con-

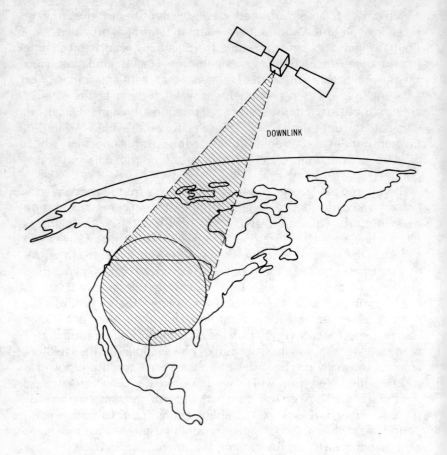

DOWNLINK

Fig. 2-12. Satellite footprint.

tours are a function of the transmitting antenna's radiation pattern and can be used to calculate receiver performance specifications which will be required for good signal reception. These types of calculations will be described in detail in Chapter 3.

As we have seen, the communications satellite is a very complex but interesting piece of technology. It combines communications electronics, rocketry, solar-energy technology, and computer gadgetry all contained within a very small chassis located at a distance of 22,300 miles above the surface of the earth. It is expected to operate flawlessly in a very rigorous environment for a period of 8 to 10 years. As a result, it is required to carry enough fuel to allow the ground station tracking facility to periodically reposition the satellite in order to correct any orbital deviations. Also, we have seen that due to space restrictions on

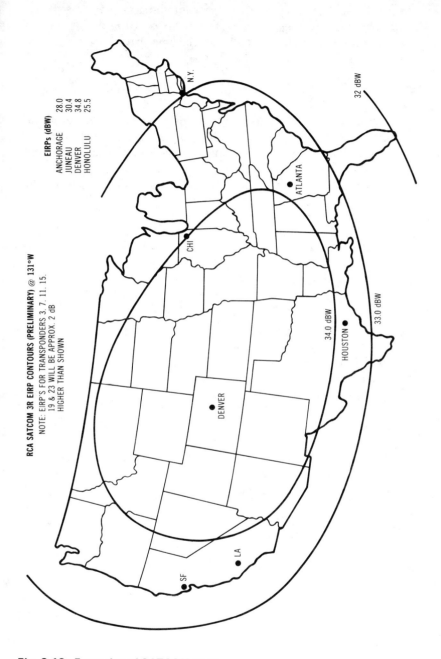

Fig. 2-13. Footprint of SATCOM III-R.

55

the satellite itself, the power radiated from the satellite for the signal's return trip to earth is very low in comparison to its uplink counterpart. This fact makes reception of such a signal a nontrivial task.

THE DOWNLINK

The characteristics of the downlink propagation path are almost identical to those of the uplink. It, too, experiences losses due to atmospheric attenuation, space loss, and antenna polarization efficiency. The only real differences between the two paths are their frequencies of operation and the fact that the atmospheric attenuation over the two paths is different. As we have already studied, the frequency of the uplink path is in the 6-GHz range, while the frequency of the downlink path is in the 4-GHz range. Because of these different frequencies of operation, the atmospheric attenuation between these two paths is somewhat different. This is true because the molecular absorption of the signal into the atmosphere varies with the wavelength of the transmitted signal. It also varies depending upon the moisture content of the air. For example, with all other things being equal, meaning that the transmitting and receiving earth stations are located equidistant from the satellite and that both are located in similar but very dry atmospheric environments, the difference in atmospheric attenuation between the two paths may be as low as 0.1 dB. If, however, the two earth stations are located in very dense rain, the difference in atmospheric attenuation between the two paths may be as great as 1 dB. In both of these cases, the uplink (6 GHz) signal is the signal which suffers the most. Keep in mind here that differences in uplink and downlink path distances and differences in their respective atmospheric environments will determine which path ultimately has the most atmospheric attenuation.

As with the uplink path, or any other RF propagation path, space loss (spreading loss) is also an inherent characteristic of the downlink path. As we have already seen, this type of loss is primarily dependent upon the distance between the satellite and the earth station with which it is interacting. It is caused by the spreading of the wavefront of the RF signal as it reaches further and further away from its source. Space loss for the downlink path is also on the order of 200 dB and is the single most devastating source of loss of the signal we are trying to receive. A loss of 200 dB to a 2500-watt signal produces a signal at the receiving antenna of only 0.000000000000000025 watt, which means that our receiving earth station must be capable of pulling an extremely small signal level out of the noise and amplifying that signal to a level suitable for processing and display on a standard television.

The polarization efficiency of the downlink propagation path is also an important concern. As with the uplink path, the polarizations of the transmitted signal and the receiving antenna should match. This should especially concern the potential owner of a TVRO (television receive only) earth station, because the polarization of the receiving antenna is dependent upon the orientation of the antenna's feed with respect to the received signal. This orientation is, of course, determined by how accurately the feed is mounted to the dish, and by how well the dish itself is attached to its mount. These are subjects we will examine in Chapter 3.

SUMMARY

The satellite TV system is a remarkable technological achievement making use of many vast and intricate technologies. It is comprised of: The satellite TV broadcast facility, which beams the signal up to the satellite in a very-narrow high-power beam in the frequency range from 5.925 to 6.425 GHz; the uplink and downlink propagation paths, which exhibit very high but definable losses to the signal traversing both directions; the satellite itself, whose main purpose is to translate the 6-GHz uplink signal to its new downlink frequency of 3.7 to 4.2 GHz; and finally, the satellite TVRO earth station, which we will examine in depth in Chapters 3 and 4. You should now have a very firm grasp on just what satellite television is all about. This background material should provide you with the knowledge you need to continue with further in-depth study of the equipment required to set up your own TVRO earth station.

Chapter 3

Receiving Antennas and Low-Noise Amplifiers

Now that we have studied the history and the overall operation of the satellite communications system, it's time to turn our thoughts to the receiving equipment that will be required if we are to receive television broadcasts "over-the-satellite." This chapter will concern itself with those components of the receiving earth station that are located in the signal path between the satellite and the receiver itself; that is, the antenna, feed, LNA or LNC, an external converter, and the coaxial cable leading into the home. As we shall soon see, these components can potentially be the most critical in the entire communications link.

First, we will examine the primary reasons behind the need for a parabolic or "dish" antenna. Then, we will attempt to describe the anatomy of a typical dish antenna and how it works, and examine a few different types of dishes and their mounts. Then we will take an in-depth look at the LNA/LNC and all of its variations. In addition, the trade-off between antenna size and LNA noise temperature will be studied. And finally, for those installations that do not have an LNC, we will examine the external converter that must be used with most receivers to convert the 4-GHz signal to a new frequency the receiver can handle.

WHY A DISH?

In Chapter 2, we alluded to the fact that only a very minute portion of the television signal transmitted from the satellite makes it to earth. This is because the satellite delivers approximately 5 watts of output signal power to the antenna, and beams the signal to a very large geographical area. Because the signal level is so low when it finally reaches the earth station, the antenna must be of a special design in order to provide very high gain and directivity. Without a dish antenna, it simply isn't possible to pull the signals out of the ambient terrestrial noise to produce a viewable picture. The dish's ability to pull these signals out of the noise is made possible by its unique construction.

Both high gain and high directivity for an earth station antenna are very important requirements for optimum system operation. The gain of a receiving antenna, similar to its transmitting counterpart, is measured in dB and is a function of the size, shape, and physical construction of the antenna. Typically, the larger an antenna, the higher is its gain for a particular frequency of operation. Of course, larger antennas typically cost more, so there will always be a trade-off. Later in this chapter, other types of trade-offs will be examined in more detail as you become more familiar with some of the systems aspects of the satellite TVRO station.

Table 3-1 will give you an indication of just how much gain you can expect from some of the various size dishes currently available. The values shown are not meant as absolute numbers but are shown only for comparative purposes. They are, however, excellent gain specifications for the intended market. Note that the gain of an antenna does not necessarily increase linearly with its size. Other factors, such as the type of feed and the antenna's RF smoothness, contribute to the overall efficiency of the antenna and thus to its gain. Therefore, just because an antenna is a certain size, do not assume that it must have a specific gain. The chances are that such an assumption could be very costly.

Table 3-1. Antenna Diameter Versus Gain

Antenna Diameter		Gain (dBi)
3 meters	(9.8 ft)	39.5
4.6 meters	(15.1 ft)	43.5
5 meters	(16.4 ft)	44.5
7 meters	(23 ft)	47.5
10 meters	(32.8 ft)	50.85
11 meters	(36.1 ft)	52

Gain Versus Smoothness

The RF smoothness of an antenna is one characteristic that is not often considered when shopping for a reliable system. We all know that for any object to reflect light efficiently it requires an extremely smooth surface. The smoother that surface is, the more light the object can reflect, and thus we say that its reflection efficiency is very high. Conversely, rough surfaces do not reflect light well at all, hence such surfaces have not found their way into the mirror industry. The interesting point of all of this is that light behaves much like any high-frequency RF signal. In physics, this is referred to as the "wave-like" nature of light. This type of behavior for light makes it very convenient to explain the operation of the dish antenna in general and, in particular, why the dish must "look" smooth to the RF signal. If the conductive surface of the dish is rough, grainy, dented, or otherwise deformed, it simply is not able to focus the RF energy into the feed efficiently. Instead, the RF energy striking the dish is reflected off in some random direction and never makes it to the feed. Thus, the efficiency of the antenna is decreased and, hence, its gain must be diminished. Keep in mind, however, that it is the conductive (metal) area of the dish that the RF energy must see as being smooth.

Most manufacturers coat or otherwise conceal the conductive surface area of the dish antenna in a protective coating such as fiberglass or plastic, thereby making it impossible to inspect its RF smoothness. If the protective coating itself is very rough, that does not necessarily mean that the conductive surface must be rough. In fact, there are many very good antennas on the market that feel rough to the touch, but which offer very good RF integrity. Antennas that are noticeably dented, however, are another story and have probably suffered some loss in reflective efficiency. The caveat here is simply this: Be sure to check the gain and efficiency specifications for the antenna you intend to buy. Do not assume that all antennas of a given diameter will perform equally well. It simply isn't true. And if you already own an antenna, strive to keep it as free from dents as is humanly possible.

Directivity

The directivity of an antenna, otherwise known as its *beamwidth*, is another very important characteristic of a parabolic antenna. It is a measure of the gain of the antenna versus the angle of deviation from its centerline projected out into space. This specification is usually given as the 3-dB and the 15-dB beamwidths of the antenna. It can be somewhat likened to the selectivity specification of a receiver, as it gives you an indication of how well the antenna can select the desired station (satellite) while rejecting all others at which the antenna is not aimed. This is an important consideration in light of the fact that all of the satellites carrying TVRO programming are transmitting on the same frequencies. Not only that, but the satellites are all beaming their signals to much of the same geographical area. Keep in mind, too, that the satellites are angularly spaced fairly close together (4°) in geosynchronous orbit.

One of the primary purposes of your TVRO antenna is to "select" one satellite at a time for your viewing pleasure while ignoring all others. In fact, the only adjacent satellite selectivity you can possibly get (other than orthogonal polarizations) is with the beamwidth or selectivity of your antenna. The narrower the beamwidth of the antenna, the less chance you will have that unwelcomed interference from adjacent satellites will invade your system. The FCC is presently evaluating the possibility of a future orbital spacing of only 2° between satellites. Such orbital spacings could render some of the presently used antennas useless simply due to an insufficiently small beamwidth. In a situation such as this, for example, an antenna with a 3-dB beamwidth of only 2° could only provide 3 dB of attenuation to a signal from an adjacent satellite. Unless the desired and undesired signals were orthogonally polarized, the signals from the two adjacent satellites would, in all likelihood, completely obliterate the desired signal, thus creating an impossible viewing situation.

Orthogonal polarizations on adjacent satellites would, of course, help somewhat due to the nature of a polarized feed in accepting only the correctly polarized signal. However, in such situations the polarization efficiency of the receiving antenna becomes extremely critical. This topic was discussed in more detail in Chapter 2. The fact remains, however, that adjacent satellites are not the only potential source of interference to your TVRO system. Telephone companies do an adequate job of creating potential interference problems with their terrestrial microwave communications systems, which also operate in the 4-GHz region. In addition, other sources of terrestrial noise (thermal and man-made) tend to find their way into the TVRO system. An antenna with a very narrow beamwidth (very high directivity) and very low sidelobes tends to minimize the influx of such unwanted signals into the system. Table 3-2 will give you an indication of the various beamwidths available for various sizes of antennas. Note that, in general, larger antennas will provide narrower beamwidths at a given frequency of operation. Again, this tends to indicate that, all else being equal, the larger your antenna is, the better off you will be.

Table 3-2. Antenna Diameter Versus Beamwidth

Antenna Diameter		Beamwidth
3 meters	(9.8 ft)	1.7°
4.6 meters	(15.1 ft)	1.12°
5 meters	(16.4 ft)	0.86°
7 meters	(23 ft)	0.7°
10 meters	(32.8 ft)	0.46°
11 meters	(36.1 ft)	0.4°

Fig. 3-1 is a graph that will provide you with a better feel for the beamwidth of a typical 3-meter dish antenna and exactly what that beamwidth can do for you. Some of you may recognize the response as being very similar to that of a bandpass filter, which is used in various capacities to pass a certain range of frequencies while attenuating all others. This is the type of filter, for example, that will provide the selectivity in a communications receiver. Note, however, that the selectivity provided is not based on frequency, but on the angular distance that you move away from the centerline of the antenna. For example, as Fig. 3-1 shows, this particular antenna will provide 18 dB or more attenuation (relative to the point at which the antenna is aimed) to any signal perceived to be emanating from a point at least 4° on either side of centerline. Thus, this antenna's 18-dB beamwidth can be said to be approximately 8° (2 × 4). According to the manufacturer, this particular antenna provides a 3-dB beamwidth of 1.7° and a 15-dB beamwidth of 3.6°; not bad specifications for a 3-meter dish.

THE ANATOMY OF A DISH

Fig. 3-2 is a very simple drawing which shows the typical geometric shape of a parabolic dish antenna. As with most any other antenna, the gain and directivity of the dish are determined by its shape and by its mechanical precision. With most other high-gain antennas, however, their gain is a function of the number of antenna "elements" in the design (see Fig. 3-3). Most antennas of this type have one or more "active" or "driven" elements (those connected to the antenna's feed-line) along with several "directive" elements (those in front of the driven element) and several "reflective" elements (those behind the driven element). This type of antenna is known as a *multielement yagi* or *beam* and is typically used for local TV reception, ham radio, and other applications. Though these antennas can be made with quite a large amount of gain and directivity, they do not even approach the possibilities available with the dish antenna. This is because the dish was designed to focus the RF energy from the satellite at which the antenna is aimed into a very intense radio-frequency beam at the antenna's focal point (Fig. 3-2). This is done in much the same manner as when a magnifying glass is used to focus the light from the sun into an intense beam that can easily be used to start a fire. In fact, if a dish antenna were made out of an extremely light-reflective material such as a mirror, and if it were aimed directly at the sun, it would produce an intense spot of light at its focal point which could easily be used to start a fire. In fact, one company markets a very small version of such a reflective dish as a solar cigarette lighter.

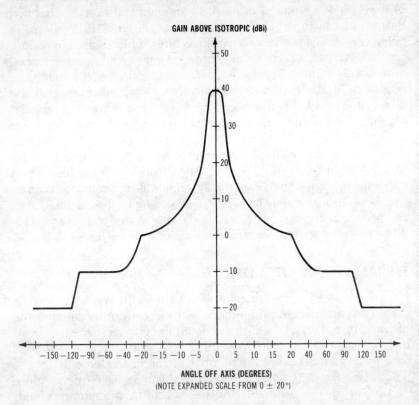

GAIN ABOVE ISOTROPIC (dBi)

ANGLE OFF AXIS (DEGREES)
(NOTE EXPANDED SCALE FROM 0 ± 20°)

Fig. 3-1. Antenna gain versus angle for a typical antenna.

Note in Fig. 3-2 that the focal length and hence the focal point of the parabola are determined by the degree of curvature and the diameter of the dish. Once the diameter (d) and the depth (h) of the dish are known, its focal length can be very easily found. Thus, if you were lucky enough to find a surplus dish on the market, you should have little

64

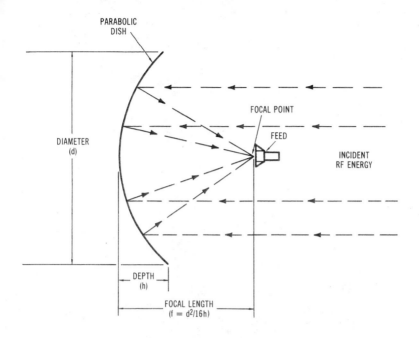

Fig. 3-2. Geometry of a prime-focus antenna.

trouble in determining its focal point by using the simple geometric relationships shown. For example, let's say that out of some stroke of luck you were able to find a surplus 10-foot diameter parabolic dish in very good condition. The only problem with the dish is that it has no mounting mechanism for the feed/LNA assembly, which means that you must design and build one yourself. This being the case, one of the very first things that must be done is to determine the focal length for the dish so that a suitable mount can be designed to place the feed precisely at the focal point. One very easy method of determining the focal point would be to lay the antenna down on a flat surface as if it were a very large bowl. Then, measure the distance from the flat surface to the edge of the dish. This measurement is indicated by the depth (h) of the parabola shown in Fig. 3-2. If this measured dimension for our hypothetical dish were 1.25 feet, then, using the equation, $f = d^2/16h$, the focal point can be calculated to be 5 feet from the center of the dish. Thus, the mount could then be designed to place the input of the feed/LNA assembly at the antenna's focal point. Fig. 3-4 shows the anatomy of a typical TVRO antenna installation.

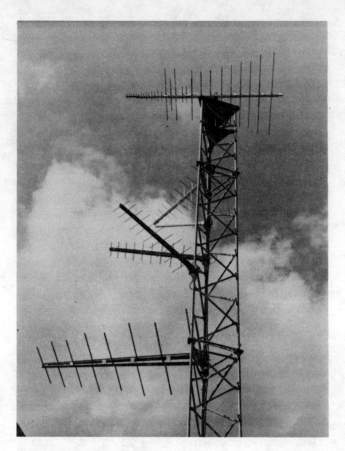

Fig. 3-3. Multiple-element yagi antennas.

TYPES OF DISH ANTENNAS

There are, of course, other shapes and sizes of dish antennas currently available on the market. Though all operate by focusing incident RF energy into an intense beam for collection by a feed, not all accomplish that task in exactly the same manner. Let's take a look at a few of the many different types of antennas available today and examine their geometry, their advantages, and their disadvantages.

The Prime-Focus Antenna

Examples of the prime-focus antenna are shown in Figs. 3-4 and 3-5. This antenna is probably the most widely used TVRO antenna in existence today. The prime-focus antenna gets its name from the fact that the feed is placed directly at the focal point of the antenna. Its operation is exactly as was described in previous sections. That is, the RF signal from the satellite, which is incident upon the dish, is focused directly into the feed as shown in Fig. 3-2. The primary reasons for its popularity are that it is relatively inexpensive and convenient for both the manufacturer and the consumer to build and use.

Fig. 3-4. Prime-focus parabolic dish antenna.

The fact that the antenna is relatively easy to build has certainly spurred a multitude of "garage-shop" operations to help fill the need in the marketplace. This is not to say, however, that all of the antennas produced by the various manufacturers are equally good. They certainly are not. However, many consumers do not know the difference between a good antenna and a bad one. Many small "garage-shop" operations, as they have been called, do provide a quality product at a very good price. As with any new industry, however, there are some that are out to make a "fast buck" without regard for quality. The caveat here is simply this: Compare quality, price, and specifications, for any antenna, especially the prime-focus type, before you buy. If the manufacturer is honest about the performance of his product, then a quick look at the spec sheet should provide you with the data you need to compare antennas. Compare the gain, efficiency, and directivity specifications of several antennas from different manufacturers before

Fig. 3-5. Another prime-focus antenna.

you buy. You will be surprised at the results and may even find a few manufacturers who will refuse to supply the information. Stay away from these companies. Tables 3-1 and 3-2 should provide you with some benchmarks for comparison of specifications to a quality product.

Though the prime-focus antenna is easy to build and is convenient for the consumer, it is not without its problems. One of the most pressing problems in any antenna design for a satellite communications application is a reduction in the amount of thermal noise the antenna receives or causes within the TVRO system. This additive extraneous thermal noise is referred to as the *noise temperature* of the antenna. The prime-focus antenna has a relatively poor noise temperature when compared to other types of antennas because of the direction that the LNA feed must face in order to collect the focused RF energy from the dish. A graphical representation of the problem is shown in Fig. 3-6. The problem lies in the misalignment of the aperture of the feed with respect to the outside circumference of the dish. The ideal condition is shown in Fig. 3-6A. In this case, the aperture of the feed exactly aligns with the outside circumference of the dish. Aperture can be likened to the peripheral vision of the feed. That is, the aperture of a feed is defined by what the feed actually "sees" in the direction it faces. Obviously, for the most efficient operation possible, it would be best if the feed were illuminated by the entire dish, as shown in the diagram (Fig. 3-6A). If the feed could only see a small portion of the dish (Fig. 3-6B), its efficiency would be poor and its gain would suffer. If, on the other hand, the opposite condition existed, as shown in Fig. 3-6C, where the feed is positioned such that it is able to see past the outside edge of the dish, another problem can be noted. This problem, often called *spillover*, causes an increase in the amount of thermal noise that invades the system. The thermal noise is added to the system by the relatively "hot" (noisy) earth in the feed's field of view. Spillover is an unwanted condition because it degrades the signal-to-noise performance of the TVRO earth station. The Cassegrain antenna, described in the next section, was an attempt to eliminate this problem.

The Cassegrain Antenna

The Cassegrain antenna design was originally lifted from the telescope design of William Cassegrain. One of the purposes of such a design was to alleviate some of the noise temperature degradation which may have been inflicted on a system due to the aperture misalignment and spillover of the prime-focus antenna. The Cassegrain antenna is a dual-reflector system as shown in Figs. 3-7 and 3-8. It employs a main reflector that is paraboloidal in shape, similar to the prime-focus antenna, and it also employs a smaller hyperbolic subreflector. Though there are variations on the dual-reflector theme, such as with ellipsoidal and other custom-shaped subreflectors to realize specific gain distributions, their operation is basically the same. Let's consider the geometry of operation shown in Fig. 3-7.

With a Cassegrain antenna, any RF wave incident upon the main parabolic reflector is directed toward the focal point of that reflector in the same manner as was described for the prime-focus antenna. Before

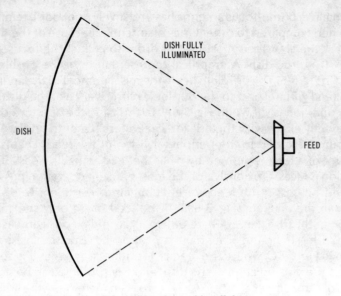

(A) Ideal condition—feed fully illuminated by dish.

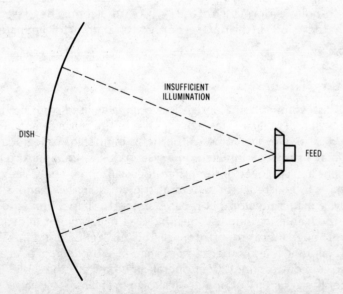

(B) Feed only partially illuminated by dish, causing reduced gain.

Fig. 3-6. Aperture alignment problem

reaching this focal point, however, the RF energy is redirected toward another focal point by a smaller subreflector system. The new focal point is located very close to the geometric center of the main dish. It is at this focal point that the feed is located to collect the focused RF energy for application to the LNA and associated electronics.

Cassegrain antennas are not often seen in home earth-station use for several reasons. They are mentioned here in order to brief you on what is available, although it is doubtful that many consumers own one or will buy one. The primary reasons for its failure to enter the home market are that it is usually larger than the prime-focus antenna, it is more critical to manufacture, and it typically costs more.

The fact that the Cassegrain antenna must usually be larger than the prime-focus antenna, especially for the market this book is addressing,

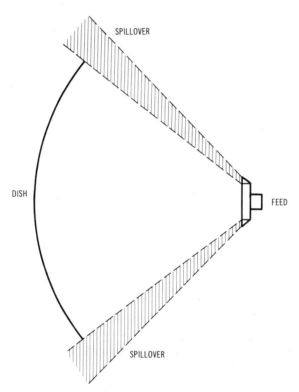

(C) RF spillover when aperture alignment allows feed to see past edge of dish, causing increase in thermal noise.

with the prime-focus antenna.

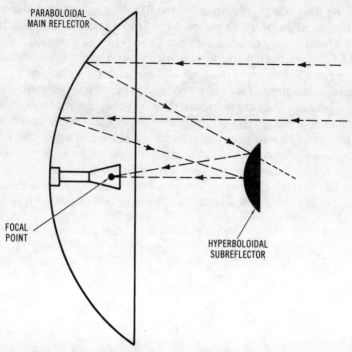

PARABOLOIDAL
MAIN REFLECTOR

FOCAL
POINT

HYPERBOLOIDAL
SUBREFLECTOR

Fig. 3-7. Geometry of a Cassegrain antenna.

is because the subreflector tends to block a significant portion of the RF energy that would normally be incident upon the main parabolic reflector. Since this energy is blocked from reaching the main reflector, it can never be directed toward the feed system via the "back" of the subreflector. Instead, the energy is reflected back into space to serve no useful purpose. This loss of energy is shown in Fig. 3-9 and is called *aperture blockage*. Aperture blockage can usually be compensated for by increasing the physical size of the main dish which, as we have already seen, typically increases the gain of the antenna. Aperture blockage isn't really a problem with large antennas (15 feet or greater) because any loss of RF energy due to such blockage can usually be adequately made up for through increased efficiency and directivity. Typically, antennas in the 10-foot category cannot tolerate the significant aperture blockage required (10% to 20%).

One of the methods that manufacturers use to improve the efficiency of the Cassegrain antenna and to overcome this significant loss of aperture is to uniquely shape both the main reflector and the subreflector such that they are still surfaces of revolution, but they are no longer a paraboloid and a hyperboloid. In other words, each of the

72

reflectors is given a unique shape that is usually derived by a computer-aided design technique. These computer techniques solve several simultaneous equations which are used to define the optimum redirection of RF energy to various portions of the two reflectors that may be inadequately illuminated due to the RF blockage problem. These types of antennas are called *dual-shaped Cassegrain* antennas and are in very wide use in commercial installations. Typically, a dual-shaped Cassegrain is capable of efficiencies approaching 65% to 80%.

The Spherical Antenna

Probably one of the most popular antennas for the home market in recent years has been the spherical antenna shown in Figs. 3-10 and 3-11. This antenna has been especially popular with the "do-it-

Fig. 3-8. A Cassegrain antenna.

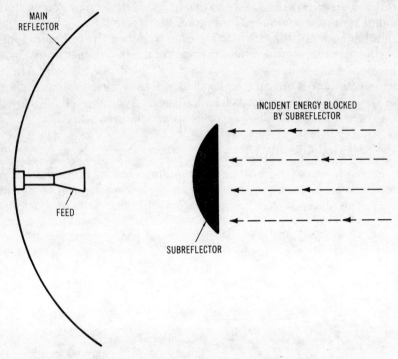

Fig. 3-9. Aperture blockage of a Cassegrain antenna.

yourselfer" because it is much easier to build than the others that have been discussed. On the other hand, the antenna does have a few disadvantages that have kept it out of the running as a low-cost alternative to the prime-focus antenna.

The geometry of the spherical antenna is somewhat similar to that of the prime-focus antenna. At first glance, the antenna looks as if it is a very flat plate with hardly any curvature. A closer look, however, will reveal a very slight curvature which can be likened to a parabolic antenna with a very long focal length. In fact, the focal length of the spherical antenna is extremely long (12 to 20 feet) when compared to a similar sized prime-focus antenna and, as such, it is one of the disadvantages of such an antenna. The curvature of the spherical antenna can be described as a section of an extremely large ball or sphere (thus the name *spherical* antenna). That is to say that if you extended the curvature of a spherical antenna for a complete revolution, you would actually trace a sphere as shown in Fig. 3-12. Like the parabolic prime-focus geometry, the spherical antenna is able to reflect RF energy back into a focal point where the feed/LNA assembly is located. And unlike

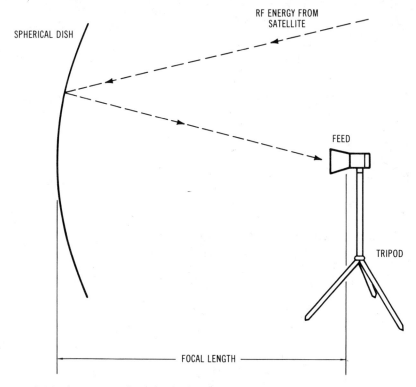

Fig. 3-10. Geometry of a spherical antenna.

the Cassegrain, this is done without a subreflector. The spherical antenna, however, has several distinct focal points that may be used to receive signals from several satellites simultaneously, providing, of course, you have enough money to spend on several feed/LNA assemblies. This type of operation is shown in Fig. 3-14. Of course, due to the direction the feed is facing, the spherical antenna may also suffer from noise temperature degradation, just as does the prime-focus antenna, due to the feed's aperture spillover.

The long focal length of a spherical antenna is shown in Fig. 3-13. As shown, the feed/LNA assembly for a spherical antenna is typically located on a tripod, similar in appearance to a camera tripod. In this case, rather than aiming the antenna at a particular satellite, the antenna is initially aimed "somewhere along the Clarke belt" and then fixed in this position. In other words, it is usually aimed at a point somewhere in the middle of a small arc of satellites that the user may be interested in. Since the curvature of the antenna is so slight, it is able to

Fig. 3-11. A spherical antenna and its feed.

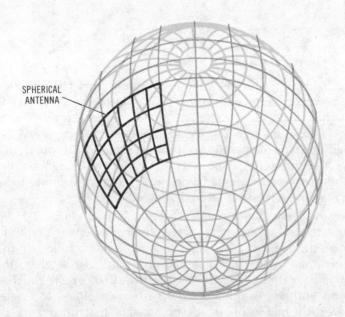

SPHERICAL
ANTENNA

Fig. 3-12. Spherical antenna as a section of a sphere.

76

Fig. 3-13. Focal length of a spherical antenna.

"see" an arc of the Clarke belt approximately 40° wide. Once aimed at the arc, individual satellites may be selected through judicious placement of the tripod/feed/LNA assembly. RF energy from every geostationary satellite within the antenna's 40° window would then be focused at a different point out in front of the reflector. A satellite positioned directly in front of the antenna would have a focal point directly in front of the center of the antenna. A satellite located in an easterly direction along the antenna's 40° window would have a focal point slightly west of center for the antenna. Conversely, a satellite located in a westerly direction along the antenna's arc would have a focal point slightly east of the center of the antenna. Switching from satellite to satellite is, therefore, simply a matter of moving the tripod a few feet in one direction or another in order to locate the feed at the next satellite's focal point.

It should be pointed out here that repositioning of the tripod may not be such a trivial task, especially if the ground on which the tripod is mounted is not flat with respect to the plane of the antenna. If there were some slope to the ground, for instance, the tripod height would have to be adjusted in addition to its lateral position. In addition, any tilt of the feed with respect to the polarization of the input signal would cause a decrease in the polarization efficiency of the antenna and would decrease the signal level available to the receiver. Therefore, any slope or tilt of the feed due to the roughness of the terrain must be compensated for with the legs of the tripod. Rough terrain can some-

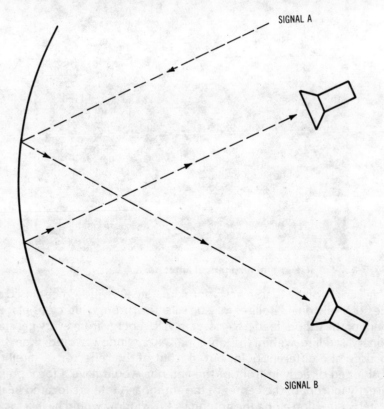

SIGNAL A

SIGNAL B

Fig. 3-14. Multiple feeds for a spherical antenna.

times make it very difficult to "find" a satellite with a spherical antenna simply because the focal point of the antenna can easily be misjudged under such circumstances. In addition, and for obvious reasons, the long focal length of the spherical antenna can cause problems in small backyards.

Though it is difficult to compare a spherical antenna to a parabolic prime-focus antenna, there is a rule of thumb that does hold true. A well-made spherical antenna with a diameter of 10 feet will typically provide the same gain as a 12-foot, 55% efficient prime-focus antenna. Or, to put it another way, a spherical antenna will typically provide more gain than the same size parabolic antenna. Of course, this will only hold true if you compare apples to apples (well-made sphericals to well-made parabolics). Though the gain may seem like a good advantage to have with a spherical antenna, the hidden "gotcha" is the fact that the sidelobe performance of the antenna is very poor. This means that the antenna is very prone to unwanted terrestrial (and extraterrestrial) interference. And for similar reasons, the noise temperature of the antenna is typically worse. Keep in mind here that these comparisons are difficult to make simply due to the various qualities of antennas available on the market today. A poorly made spherical antenna will perform no better than a poorly made parabolic.

Though the spherical antenna has been extremely popular with the hobbyist in the infancy of satellite TV reception, the prime-focus antenna, due to its simplicity and convenience of operation, has become the mainstay of the home satellite antenna. Though the spherical antenna can be more easily constructed by a hobbyist, the parabolic antenna lends itself well to mass production and is less difficult to use than a spherical due to its self-contained focal point and its good adjacent satellite rejection.

The spherical antenna has often been advertised as a "build-it-yourself" type antenna. The frame can be made from wood and the surface of the antenna can be fashioned from wire mesh or screen. Keep in mind, however, that wood is subject to warping when exposed to the elements. Thus, these antennas will not maintain good surface tolerance when used outside, especially in icy, windy, or wet climates.

SUMMARY—ANTENNAS

The satellite TV antenna is a very important consideration for any earth station. Its purpose is really twofold. First of all, the antenna is used to provide quite a large amount of signal gain to the 4-GHz information incident upon the surface of the dish. It does this by focusing the RF energy into an intense beam at the antenna's feed. Second, the antenna is used to discriminate against unwanted satellite or terrestrial signals that may be operating in the same frequency range. It does this through its narrow beamwidth for a given focal point.

Never assume that all antennas of a given size or shape are equivalent. Check their specifications and you may be surprised. Especially useful specifications to know for an antenna are its gain, beamwidth, and efficiency over the entire range of operation (3.7 to 4.2 GHz).

Later in this chapter, after we study a little about the LNA and begin to understand its function, we will begin to focus on a few of the games you can play through systems calculations to trade-off antenna gain for LNA noise figure. Such trade-offs may come in handy someday if good performance for one or the other components is difficult to find at a good price. Sometimes, a good LNA can make up for a poor dish and vice versa.

THE ANTENNA MOUNT

There are several different mounting methods for dish antennas on the market today. Though all serve a useful purpose, each does have its advantages and disadvantages. Probably the two most popular mounts are the polar mount and the AZ-EL (azimuth and elevation) mount. The polar mount is the most popular for the home earth-station market and the AZ-EL mount is most popular with the commercial marketplace. We will explore the reasons for this apparent market split in the paragraphs that follow. The antenna mount is the mechanism which helps to aim the antenna toward the satellite of interest, while providing the structural integrity of the antenna's mechanical interface with the earth's surface.

The AZ-EL Mount

One of the most familiar methods of aiming anything at any other object is the azimuth and elevation method. Without ever realizing it, you use this method every day whenever you point or even look at another object. For example, when you point at another object across the room with your finger, you align your finger with the object both horizontally and vertically. The horizontal direction is called *azimuth* and the vertical direction is called *elevation*. Those of you who are marksmen or hunters know that aiming at a target that is very far away is extremely difficult, and it requires much greater skill and accuracy than pointing with your finger at an object across the room. Aiming at a satellite 22,300 miles away is no easy task either, especially when you can't even see the target. Because you can't see the satellites, finding one with your new dish can be quite a chore if you don't know where to look. That is the reason the satellite locations, usually given in azimuth and elevation coordinates, are published in various trade publications. Additional information on how to find the satellites is given in Appendix C.

The AZ-EL mount, shown in Fig. 3-15, provides you with the capability of aiming your antenna at a satellite by moving it in two directions, horizontally and vertically. Each movement is typically done manually with the aid of a hand crank. One critical aspect of any type of aiming mechanism is that it must, of course, be precise. A good measure of just how precise an AZ-EL mounting mechanism is would be to measure the change in position of one axis as the antenna is swept over a

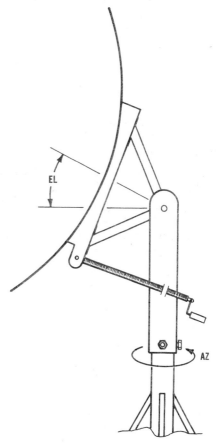

Fig. 3-15. The AZ-EL mount.

complete arc of the other axis. If, for example, you were to sweep your antenna from left to right, or right to left (azimuth), as far as it would go, the angle of elevation for the antenna should not change at all. Similarly, an elevation sweep of the antenna should produce no movement in azimuth. This type of mechanical precision, however, is not really possible and some slight variation from perfect will occur. This is true for two reasons. First, in order for such a perfect sweep of an antenna to occur, the mount for the dish must be mounted perfectly level on the surface of the earth. For obvious reasons, a perfectly level surface on which to attach the antenna's mount is not possible. There will always be some degree of error in the smoothness of such a surface. Second, the mount's own mechanical precision enters into the picture. An excellent antenna mount, if it is mounted within $1/8$ inch of being level on the surface of the earth, will produce an elevation pointing error of 0.1° or less for a 50° change in azimuth position. Though there are antenna mounts on the market which produce less than favorable results when a similar test is conducted on them, with most of the mounts currently on the market, the majority of the pointing inaccuracies that occur on one axis when the other axis is swept are due to inadequate surface preparation and poor leveling of the mount. Thus, the more time that is spent leveling an AZ-EL mount prior to antenna installation, the less time it will take you to aim your antenna at the satellite of interest after the antenna has once been installed.

The AZ-EL mount is typically used in installations that are fixed on one satellite and are not moved to look at other satellites very often. Thus, most cable TV companies will own AZ-EL mounts for their antennas. Such a commercial installation would typically have several dishes with each pointed at a different satellite. Very rarely would they be moved. Earth stations that frequently hop from satellite to satellite, on the other hand, would find the AZ-EL mount to be very cumbersome because of the two-step aiming process that must be followed in order to find another satellite. That is, both the azimuth and elevation of the antenna must be set separately and precisely to the coordinates of the satellite you are trying to receive. You can imagine how tedious and time-consuming it would be if you simply wanted to scan the entire Clarke belt every night just to see what programming was available. It would take hours! For those who find that satellite hopping is the norm rather than an extreme rarity, there exists another type of mount, called the *polar mount*, which provides an easier method of aiming at each satellite, and an easier method of scanning the entire Clarke belt with minimal effort.

The Polar Mount

In order to understand the beauty of the polar mount and exactly what it does for you, it is first necessary to understand how the Clarke belt is positioned in space relative to a typical earth station here in the United States. This relationship is shown in an exaggerated scale in Fig. 3-16. If you were able to see each satellite in the Clarke belt with the naked eye, they would appear to trace an arc in the sky due south of your position, as shown in the drawing. That is the reason that both azimuth and elevation for an antenna must be changed in order to hop from one satellite to another. Each satellite is at a different elevation angle with respect to the antenna. But note the definition of the arc. The satellites are not "all over the sky," but instead are positioned on a very well-defined permanent curve. The polar mount takes advantage of this well-defined arc by automatically forcing the elevation angle of the antenna to follow the arc as the antenna is swept in azimuth.

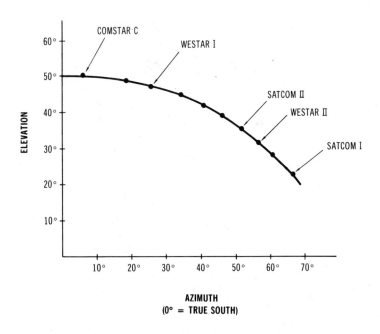

Fig. 3-16. The polar arc of satellites as viewed from the south-eastern United States.

83

The polar mount, shown in Fig. 3-17, is a mounting mechanism which allows the antenna to be aimed at a satellite, or moved from satellite to satellite, with a single step as opposed to the two-step process of the AZ-EL mount. Such a mount is configured such that when the antenna is moved in azimuth, the mount automatically changes the angle of elevation of the antenna in order to trace the arc of Fig. 3-16. Because the aiming of the elevation axis is performed mechanically, its mechanical precision must be very good. In fact, the mechanical precision of the polar mount must be much greater than

Fig. 3-17. The polar mount.

that of the AZ-EL mount. This is true because with the AZ-EL mount, changes in one axis that are caused by movement in another axis can be corrected manually through an iterative "tweaking" process. With the polar mount, however, the antenna must correctly sweep the entire arc without any "fine tuning" for each satellite. If the elevation angle of a correctly built polar mount were ever modified, the antenna would never again be able to sweep the entire satellite arc correctly. It might be able to see a few of the satellites within the arc, but not as many as it was originally designed to see.

Because of the precision required of the polar mount in order to trace the required arc, you can imagine the requirements for installation of the mount as far as ground surface preparation and leveling are concerned. Recall that with the AZ-EL mount, inadequate ground leveling, or inadequate leveling of the mount itself, caused changes in the elevation angle of the antenna when the antenna was swept in azimuth. Recall too, however, that the AZ-EL antenna was able to compensate for such a deviation through careful "retweaking" of the antenna's elevation angle. With the polar mount, such "retweaking" of the antenna is not possible. The antenna either follows the arc or it doesn't, and it all depends on the accuracy of the polar mechanism of the mount itself, and on precise installation of the mount at the antenna site.

Another critical requirement in installation of a polar mount is its precise positioning relative to due south. It must be initially "pointed" in the correct direction or it will never be able to trace the Clarke belt. The difficulty in correctly installing a polar mount has caused at least one mount manufacturer to call the polar mount "worthless" and a "royal pain to install." But, if you are able to withstand the pain and rigors of installation, a correctly installed polar mount could be worth its weight in gold.

THE FEED

The purpose of the antenna's feed is to efficiently couple the RF energy reflected from the dish into the LNA. This must be accomplished without allowing any excess thermal noise spillover from around the edge of the dish, while at the same time trying to adequately "illuminate" the entire dish. This is a trade-off that antenna and feed design engineers have been fighting for many years. The feed would like to see as much of the dish as is possible in order to increase the amount of signal energy that it receives and thus to increase the overall efficiency of the antenna. The problem arises from the fact that every feed has its very own "antenna pattern." That is, it too is a directional antenna and as such has its own beamwidth or directivity.

The ideal beamwidth for a feed would be one which covered, or illuminated, the entire dish with equal gain, while providing infinite attenuation of any signal at the outside edge of the dish and beyond. Since no feed is perfect, rather than providing infinite attenuation at the outside edge of the dish, a typical feed will only provide a slight amount of attenuation, thus allowing other signals and excess thermal noise into the system. The ideal-versus-typical feed sensitivity, or beamwidth, is shown in the graph of Fig. 3-18. Note that because the sensitivity of an actual feed diminishes as the edge of the dish is approached, signals reflected from the outer edges of the dish are attenuated by the feed before entering the LNA. Thus, the antenna actually looks smaller than it really is or, in other words, its efficiency is reduced. Note that if a perfect feed could be developed, antenna diameters would decrease dramatically due to the increase in efficiency. Some typical feeds are shown in Fig. 3-19.

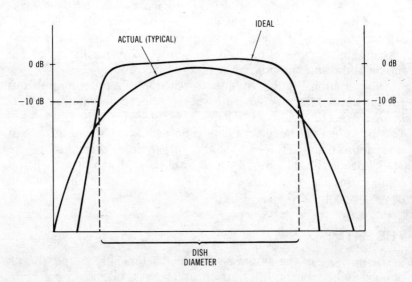

Fig. 3-18. Ideal versus typical feed sensitivity.

FEED ROTATORS

In Chapter 2 it was mentioned that both horizontally and vertically polarized signals are available for reception with a TVRO earth station. It was also mentioned that for optimum operation, the polarization of a receiving antenna must match the polarization of the received signal.

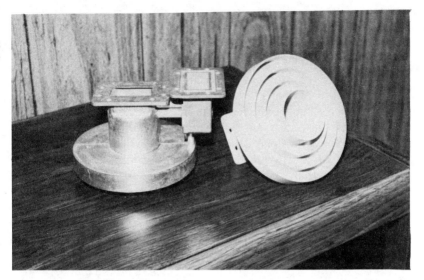

Fig. 3-19. Some typical antenna feeds.

Historically, in TVRO applications, this match has been achieved through physically rotating the entire feed/LNA assembly until it was oriented correctly with respect to the received signal. Making this adjustment manually was often an involved process which required the user to unbolt the entire assembly from its mount, rotate it, and then bolt it back again. Such a procedure is obviously very tedious. The more channel changes that were required, the more tedious the job would get.

In order to make selection of orthogonally polarized transponders much easier for the user, the motorized feed rotator was developed. This device consists of an electric motor and gear mechanism which is attached to a rotatable LNA/feed assembly. In order to change polarizations using the motorized feed rotator, the user simply toggles a switch and allows the electric motor to do the work for him. The switch is sometimes physically located outdoors at the antenna which, of course, requires that the user go outside to change polarizations. Most often, however, the rotator is equipped with a remote switch that can be located in the home in order to allow for more convenient channel selection. Some receivers even provide interfaces to feed rotators, which will automatically rotate the feed to the correct polarization for the selected transponder.

There are other types of polarization selectors on the market which do not physically rotate the entire LNA/feed assembly. Instead, they will either rotate a very small probe located inside the feed, or they will electronically alter its polarization with absolutely no moving parts. Such devices offer an advantage over the motorized feed rotator because they are physically smaller and have fewer moving parts. Typical feed rotators are shown in Fig. 3-20.

THE LOW-NOISE AMPLIFIER

The low-noise amplifier (LNA) is one of the most important components of any satellite earth station. It and the system antenna will virtually determine the "figure-of-merit" for the entire receiving system. The LNA gets its name from the fact that it provides quite a lot of gain to the extremely low-level signal present at its input without adding much noise to the signal. You will recall from earlier discussions that the signal we are trying to amplify is so low in level that it is even below the terrestrial noise floor. That, of course, is the reason why the high-gain dish antenna must be used, and is also why we have been so concerned with correctly illuminating the dish with the LNA's feed. If the feed and dish are not set up for correct illumination as per our previous discussions, the result is an increase in the amount of terrestrial thermal noise added to the signal. The object is to provide as much gain to the signal as is possible without adding any extraneous noise that may degrade the system's overall performance.

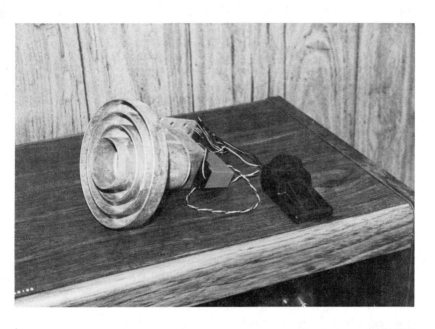

Fig. 3-20. Typical feed rotators.

89

In this section, we will look at the LNA in detail. We will examine why it's used, what it looks like, and take a brief look at a typical block diagram for such an amplifier with the purpose of providing you with an idea of the technologies involved in its design and fabrication.

An Overview

The low-noise amplifier is a device used to amplify the signal present at its input to over 100,000 times its original value, while adding hardly any noise to the signal. Though such a gain specification is certainly impressive, especially to the novice, it is really the unit's noise performance that is most important in the design of an LNA. Many different amplifiers could provide a gain specification well over 100,000 (50 dB) with no trouble at all. In fact, such amplifiers do not even come close to pushing the state-of-the-art in electronics. The design of the LNA, however, does consistently push the state-of-the-art in order to achieve better and better noise performance.

The noise performance of an amplifier is, of course, a measure of how much noise the amplifier adds to a signal as it passes through the amplifier. This performance has been specified for some amplifiers through the use of the term *noise figure*. Indeed, such is the case with the majority of lower-frequency amplifiers available today. Most will show noise figures anywhere from 3 dB to 8 dB. The LNA's noise performance, however, is usually specified with a term called *noise temperature*. Such a term is usually given for amplifiers that have a very low noise figure. The two terms are interchangeable, and it is possible to convert from noise temperature to noise figure and back again if you really find it necessary. To make it easier, however, Table 3-3 lists most of the commonly available values of noise temperature for LNAs currently on the market, along with the corresponding value of noise figure. The important point to remember here is that the LNA should add as little noise to the signal as possible. In order to do that, the LNA pushes the state-of-the-art, which is one reason that it costs as much as it does.

Table 3-3. Noise Temperature Versus Noise Figure

Noise Temperature (K)	Noise Figure (dB)
120	1.5
100	1.28
90	1.18
80	1.06

Fig. 3-21 shows the basic anatomy of an LNA. Note that the electronics is packaged in what looks to be a very expensive metal chassis. This is indeed the case. The LNA is usually mounted in a position which forces it to be exposed to the elements and, as such, it must be packaged in a weatherproof enclosure capable of withstanding extremes in temperature, humidity, ice, and rain. In addition, since the unit is mounted outside with the dish, the housing must also be RFI (radio-frequency interference) and EMI (electromagnetic interference) "tight," meaning that it must protect against RF ingress and egress. All of the above require that the housing be very well built and "leakproof." In addition, the mechanical structure must be capable of supporting its own weight plus that of any attached cables, while bolted to the feed assembly. Such strength is provided through thick metal housings and compact designs.

Fig. 3-21. Typical low-noise amplifiers.

The LNA is usually mounted directly at the antenna in order to prevent any unnecessarily long cable runs from the dish's feed to the LNA assembly itself. Due to the inherent attenuation characteristics of coaxial cable at 4 GHz, it is best to minimize any length of cable that must carry such high-frequency information, especially prior to the LNA. Cable loss prior to the LNA degrades the noise performance of the system and cannot be tolerated. There are special types of coaxial cables that have been designed primarily for transmission of 4-GHz signals, but they are expensive and still provide a large amount of attenuation at those frequencies. A later section will expand upon this brief introduction to cable characteristics. For now, let it suffice to say that the LNA should be mounted as close to the feed/antenna assembly as possible in order to minimize any length of coaxial cable or waveguide that must carry the 4-GHz RF signal.

The LNA is usually attached directly to the feed and consists of the following basic functional blocks: the waveguide input, an isoadapter, a very-high-gain transistor amplifier, and a coaxial output connector. The following paragraphs will describe each functional block in detail in order to better familiarize you with the operation of the device.

The Anatomy of an LNA

Fig. 3-22 shows a simplified block diagram of an LNA. As was mentioned previously, the LNA approaches the state-of-the-art in low-noise electronic design. For this reason it is very difficult to design a high-quality LNA at home. Proper design requires the use of sophisticated computer techniques in transistor modeling and interstage network design for optimum noise performance. There are a few kits available on the market that do offer fair performance, but these kits were most likely designed using computer techniques, and if not, they are probably not optimized as well as they could be. Remember, the primary consideration in the design of the LNA is noise temperature. In fact, other performance specifications such as gain are readily com-

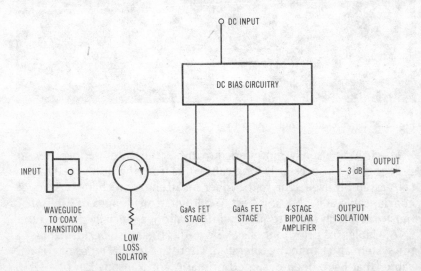

Fig. 3-22. Block diagram of a low-noise amplifier.

promised to improve the noise performance of the amplifier.

The Waveguide—At microwave frequencies, those frequencies at which the LNA operates, coaxial cable tends to have very high losses. These losses simply cannot be tolerated prior to the LNA. In addition, signal energy being reflected from the dish and into the feed is in a form (transverse electromagnetic waves) that coaxial cable cannot efficiently transport to the electronics. For these reasons, a waveguide is typically used to transport RF energy from the feed to the LNA assembly. In fact, the feed itself can be considered to be a flared waveguide. A waveguide is simply a hollow rectangular or tubular conductor that is used to guide waves of RF energy from point to point. The RF propagation path is not in the actual conductive material of the waveguide, but instead is in airspace that the conductor surrounds. Waveguides typically have much lower attenuation characteristics than a similar length of coaxial cable and, for this reason, are quite often used at microwave frequencies to transport RF. Waveguides come in all shapes and sizes depending on the frequency of operation and the application. For TVRO applications at 4 GHz, a rectangular waveguide will typically have cross-sectional dimensions of 1.15 inches by 2.3 inches. This type of waveguide is referred to as a WR229 waveguide. Typically, the LNA will have a short section of waveguide built into the unit, as shown in Fig. 3-21. This waveguide will have a flange that mates with the WR229 flange on the feed so that the two can be easily mounted together.

The Isoadapter—The isoadapter is a device that serves several different functions. Part of its function is to provide the translation from the waveguide input to the coaxial (wire) mode that is required by the electronics that follow. The theory behind such a translation is beyond the scope of this book and will not be discussed. The isoadapter also functions as an isolator which provides a good termination for the feed, and provides the LNA with a constant resistive impedance that is used to derive the first amplifier's minimum noise-figure match. The isolator is a device that passes RF energy in one direction but not in the reverse direction. It is used in this case to pass energy from the feed to the amplifier stages, but disallows RF flow from the amplifier stages out to he feed. Therefore, any signal incident upon the electronic circuitry, but reflected back toward the feed due to an impedance mismatch, is absorbed by the isolator before it makes its way back to the antenna. The antenna, because it never sees any reflected energy from the isolator, has a good termination by definition. In a similar manner, the first-stage amplifier never sees any reflected wave from the isolator and, therefore, the source impedance that it sees is constant, which is the ideal condition. Though much of the previous discussion is some-what technical, it does provide you with a general overview of what functions the isoadapter performs. To put it very simply, it is the interface between the electronics and the plumbing.

The Amplifier—Following the isolator is a gain block called the *low-noise amplifier.* This is the circuitry whose characteristics are so vital to the overall operation of most satellite earth stations. The actual gain block is typically made up of several transistor amplifier stages in cascade, with each stage being optimized for best noise performance. The first two stages of amplification are usually provided by a device called a *gallium-arsenide field-effect transistor* (GaAs FET). The GaAs FET (pronounced gasfet) is a transistor that has revolutionized low-noise circuit design as we know it today. These devices by themselves are currently capable of noise figures as low as 0.6 dB at 4.0 GHz at gains of greater than 14 dB. Once placed in production units, however, you might expect a little poorer performance in both gain and noise figure, though noise figures of 1 dB and lower are certainly possible.

Prior to the development of the GaAs FET, most LNAs were of the extremely expensive parametric variety and were totally out of the average consumer's price range. Now that the GaAs FET has been on the market for a while, its price has decreased and its performance has improved drastically. The price erosion has occurred primarily due to the increase in production volume for the device. GaAs FETs are increasingly being used in very high volume consumer electronics such as the television set. As these volumes continue to grow, we can expect the price of the GaAs FET to continue to decrease, and ulti-mately for the price of the LNA to substantially decrease also. The improvement in performance of the GaAs FET, and hence the perfor-mance improvement of the LNA, can be approximated by the curve of Fig. 3-23. Note that in late 1972, the best noise temperature that could be expected of an uncooled amplifier was on the order of 290 kelvins (K). Over the years, however, the GaAs FET has improved so dras-tically that in 1980 uncooled 80-K LNAs were a reality, and in the very near future, the 50-K LNA will be rolling down the production lines.

The term *uncooled* is one that perhaps requires some definition. The noise that a device or group of devices (such as those making up an amplifier) adds to a signal can originate from many different sources. One of the largest contributors of noise to any system is that caused by the random motion of the molecules within the device. This noise contribution is called *thermal noise* and it is directly related to the temperature of the device. The hotter the device is, the faster is the molecular motion, and the higher its noise contribution will be. On the

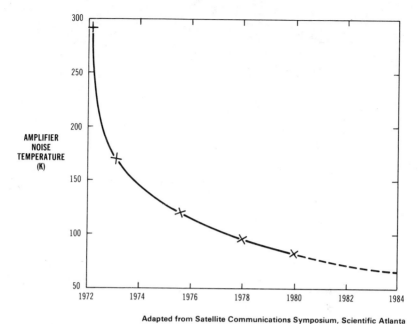

Adapted from Satellite Communications Symposium, Scientific Atlanta

Fig. 3-23. Improvement in LNA noise performance over the years.

other hand, by lowering the temperature of a device, we can slow the random motion of its molecules and hence lower its noise contribution. Theoretically, if we were to lower the temperature of a device to absolute zero (–273.15° Celsius, or zero on the Kelvin (K) scale), all molecular motion within the device would cease, and there would be no thermal noise. This concept is actually used to lower the noise figure of some extremely low-noise amplifiers in very critical applications. Such amplifiers are called *cryogenically cooled* amplifiers, and, as you can imagine, are very difficult to maintain.

Each GaAs FET stage shown in Fig. 3-22 will contribute approximately 12 dB of gain to the overall system and will virtually set the noise figure of the entire amplifier. Though each transistor is certainly capable of much more gain than 12 dB, the gain is sacrificed to improve the overall noise figure of the design. If the design were optimized for gain instead, its noise figure would suffer tremendously. Though the block diagram indicates that two GaAs FETs are placed back to back in the design, some of the lower-cost and higher noise-figure units use only one such stage prior to the more conventional bipolar transistor stages. The problem with this approach is that the gain of the first stage is not high enough to eliminate the noise contribution of the succeeding stages. Since the noise contribution of a bipolar transistor amplifier is extremely high in comparison to that of the GaAs FET amplifier, it adds to (although not directly) the contribution of the first stage. With two GaAs FET stages preceding any bipolar amplification, however, the noise contribution of the bipolar transistors is virtually eliminated, thus producing a lower-noise design.

The four stages of bipolar transistor amplification shown produce approximately 32 dB of gain to round out the amplifier design. Though these stages consist of the more conventional bipolar transistors, their design is certainly not trivial. Each stage must, of course, provide a very flat frequency response from 3.7 to 4.2 GHz with minimal gain variation through extremes in temperature. These four bipolar transistor stages are followed by a 3-dB pad, or attenuator, which is used to provide some isolation between the final transistor and its load. Such a device helps to protect the transistor from load mismatches that may occur from time to time.

Power and Bias Circuitry—The LNA typically receives its DC power from either the center conductor of the coaxial cable at its output, or through a separate connector on the housing. An internal power supply regulates these supply voltages and applies them to the circuitry throughout the amplifier. In addition, a voltage inverter may be included to transform positive supply voltages to negative voltages in order to provide bias to the gates of the GaAs FETs.

Overall Noise-Figure Performance

As was stated earlier, the overall performance of an LNA is typically gauged by its noise figure, or noise temperature. One thing that must be understood, however, is that the noise temperature of an LNA is dependent upon two very important factors—frequency and ambient temperature.

Fig. 3-24 is a graph of LNA noise figure versus frequency for two typical amplifiers. Note that the noise figure for an LNA will typically be much worse at the high end of its frequency range than at the low end. This is simply because the overall amplifier noise-figure response very closely resembles the noise response of the transistors that make up the amplifier. The specified noise temperature for the LNA should be the worst-case noise temperature across the frequency range of interest. For the case shown, the specified noise temperature should not be any better than 92 K.

It was also mentioned earlier that the noise temperature of a device is dependent upon the ambient temperature. The hotter it gets, the worse will be its thermal noise contribution and hence its noise temperature. The LNA is no exception. For this reason, the noise temperature of an LNA is always specified at room temperature. If the LNA is placed outdoors, its true noise temperature may look something like the graph of Fig. 3-25. Note that with a temperature variation of 50° Celsius, an amplifier's noise temperature may vary as much as 20 K (approximately 0.2 dB). For this reason, you should always plan for the worst-case noise temperature and design your system around that value. If, on the other hand, your earth station is already in operation, and if you happen to be operation in marginal signal conditions, you may have already seen the results of this phenomenon by an improvement in video signal-to-noise ratio during the winter, or perhaps some degradation in the summer months. The moral of the story is, of course, not to design systems around "typical" device specifications.

The LNA is a marvel in state-of-the-art electronics. Its improvements in recent years, primarily due to the GaAs FET, have brought prices down to within the grasp of the consumer while substantially improving its noise temperature performance.

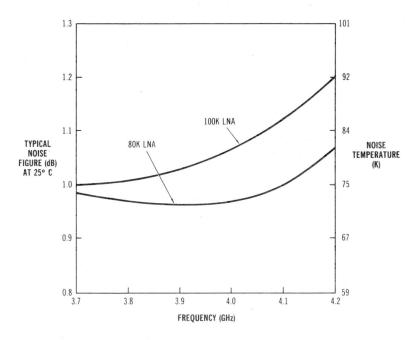

Adapted from Satellite Communications Symposium, Scientific Atlanta

Fig. 3-24. Typical noise figure versus frequency curves.

97

THE CONVERTER

After amplification through the LNA, the received satellite TV signal is still at a frequency within the 3.7- to 4.2-GHz range. Not only are such frequency signals incompatible with present-day television sets, but as we have stated before, coaxial cable losses are extremely high at these microwave frequencies. In most home installations, such losses simply cannot be tolerated, especially in long cable runs from the dish antenna into the home. For this reason, many home earth-station installations are equipped with a device (usually located at the antenna) that translates a particular transponder channel down to a lower frequency called an *IF*, or *intermediate frequency*. The signal is then transported into the home and fed to the satellite receiver via inexpensive coaxial cable. Such cable has relatively low loss at these new frequencies. The device in question is called a *frequency converter* because it converts the 3.7- to 4.2-GHz transponder channels to a new frequency, typically 70 MHz.

A simplified block diagram of a frequency converter scheme is shown in Fig. 3-26. The unit basically consists of a mixer, local oscillator

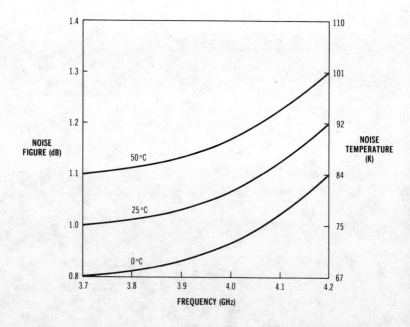

Adapted from Satellite Communications Symposium, Scientific Atlanta

Fig. 3-25. Typical LNA noise temperature versus ambient temperature curves.

(LO), IF amplifier, and a method of remotely controlling the frequency of the local oscillator. (It is by varying the frequency of the LO that channel selection is accomplished.)

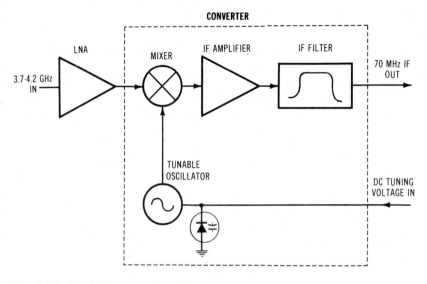

Fig. 3-26. Block diagram of a converter.

Recall from our previous reading that a satellite may have up to 24 transponders carrying video programming. Each of these transponders has its own frequency assignment within the 3.7- to 4.2-GHz downlink bandwidth. All signals present at the antenna and within this frequency range will be amplified by the LNA and, in this case, will be applied to the input of the frequency converter. At the mixer, these input frequencies are translated to a new lower frequency, which is typically 70 MHz. This is done through mixing with the local oscillator. After the mixer, the 70-MHz IF amplifier and its filtering networks are used to help eliminate all signals except for the transponder of interest. This amplifier provides some of the selectivity required and provides enough amplification to overcome any losses that may have occurred in the mixer. It also has enough gain to drive a significant length of coaxial cable and to overcome the losses the cable may induce.

With such a scheme, in order to select transponders, some method of tuning the converter's local oscillator frequency from a remote location is required. Since the frequency of the local oscillator determines the transponder that is selected, in order to change transponders, the user must be able to accurately control the frequency of the local oscillator. Because the local oscillator, along with the rest of the frequency converter, is often located on the dish's mounting structure, in order to change transponders, the LO must be remotely tunable. This is typically done via a control cable which carries a DC voltage from the satellite receiver out to the frequency converter assembly. This DC voltage is used to vary the frequency of the local oscillator. Such an oscillator is called a *voltage-controlled oscillator* (VCO).

Though there are significant advantages in the use of a converter at the dish, it does have its disadvantages. It is true that the use of an external converter does simplify the design of the receiver, and makes the resulting receiver less costly by taking out all microwave circuitry. It is also true that using an external converter simplifies and cost-reduces the coaxial cable that must be used to carry the signal from the dish into the home. But it also forces us to use an additional cable in the interface between the dish and the home—the one to carry the DC control voltage for the local oscillator. In addition, due to the selectivity of the converter's IF filter, only a signal from one transponder at a time ever reaches the home. All users in such a system must watch the same channel (transponder). Viewing several transponders simultaneously would require separate frequency converters and cable runs from the dish to each receiver in the home. That could be quite a wiring chore in a multiple-receiver installation. Due to the number of converters that would be required, it could also be very costly. For this reason, many multichannel installations are using a new frequency conversion scheme which tends to be less costly. This type of conversion scheme is called *block conversion* and is the subject of the following paragraphs.

Block Conversion—The Low-Noise Converter

Many LNA/receiver manufacturers recognized early the limitations of both the straight 3.7- to 4.2-GHz receiving systems and the frequency converter scheme discussed in the previous paragraphs . These same manufacturers are now attempting to design better conversion schemes which retain the advantages of a converter system while at the same time minimizing its disadvantages. One such scheme is called a *block conversion* approach. In the block conversion approach, instead of a single transponder being converted to a new IF, the entire 3.7- to 4.2-GHz satellite band is translated to a new frequency. This is done in a broadband mixer and broadband IF amplifier as opposed to the narrow band IF amplifier of the previously mentioned frequency converter. For example, in various block conversion techniques, the 3.7- to 4.2-GHz band might be translated to any one of the following IFs: 270 to 770 MHz, 500 to 1000 MHz, 1000 to 1500 MHz, and so on. Any

500-MHz block of frequencies is a potential candidate for a new IF frequency. Each part of the spectrum, of course, has its advantages and disadvantages and, as a result, many of the CATV equipment manufacturers are debating which part of the available spectrum is the optimum block. It is not our intent to feed the fire by promoting one block over the other. The intent here is simply to present a concept. All frequencies chosen by the various manufacturers do provide the same advantages previously mentioned for the straight converter scheme, but they also provide one additional advantage. Since the entire 500-MHz satellite bandwidth (all 24 transponders) is now on the cable, it's now possible to use multiple receivers in the home without the need for additional converters and associated cabling. In addition, since the local oscillator in the block converter is fixed in frequency, there is no need for an extra DC control cable from the receiver to the converter to control its frequency. Instead, all transponder selection is done within the receiver.

Fig. 3-27 is a block diagram of a typical low-noise converter (LNC). Note that the device is nothing more than a low-noise amplifier and a broadband converter in the same package. In fact, an LNA and an LNC are mounted in precisely the same manner on the dish, and unless you were familiar with the actual device, it would be extremely difficult to tell the difference between the two. Perhaps one noticeable difference might be the type of coaxial cable which terminates the output of the device. Hardline would most likely terminate the output of an LNA, while low-frequency coax (RG-8/U) might prevail in LNC installations. For short cable lengths at 4 GHz, however, RG-8/U is often used.

One very important thing to remember here is that the LNA or LNC must be compatible with the receiver. Many manufacturers produce LNCs that are compatible with only their receivers. This compatibility is primarily based upon the particular manufacturer's choice of intermediate frequencies. If, for example, manufacturer A decides on an LNA/converter/receiver combination that will operate with an IF of 70 MHz, you would not be able to substitute manufacturer B's receiver which is expecting a 500-MHz wide IF located between 270 and 770 MHz. It just won't work. Of course, it is possible to substitute one receiver for another if they are both expecting the same input frequency or band of frequencies.

THE ANTENNA/LNA TRADE-OFF

Thus far in this chapter, we have concerned ourselves with the task of extracting an extremely low-level signal out of the noise in order to produce a viewable picture. This requirement has forced us to look at some technologies you may have never seen before, such as the dish and the LNA. Both the dish antenna and the LNA play an equally important role in producing a quality picture. In fact, either of the two can be made to correct deficiencies in the other. In other words, within

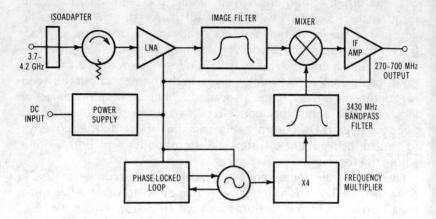

Fig. 3-27. Block diagram of a low-noise converter.

reason, a good LNA may be able to compensate for a poor antenna, and vice versa.

The gain of an antenna and the noise temperature of the LNA can be juggled without too much trouble. The larger the antenna is, the poorer the noise temperature of the LNA can be. Similarly, the smaller the antenna is, the better the noise temperature of the LNA must be in order to produce the same quality picture at the television set. This type of trade-off is often encountered when an individual is trying to put together a TVRO system from surplus components, or when buying off-the-shelf components that are not parts of a complete system. Here, the individual has the capability of maximizing performance while minimizing cost through the judicious purchase of "sale" items. The idea here, of course, would be to purchase the lowest-cost antenna/LNA combination that would still provide you with an acceptable television picture. Table 3-4 will give you an idea of the type and magnitude of trade-off which may be typical in your location. Keep in mind here that each circumstance is unique and depends on a number of different factors, such as the EIRP contour on which your TVRO station lies, the receiver noise figure, the length of cable from the dish into the home, and any other losses that may occur within the system. Table 3-4 will give you an idea of what the trade-off entails.

102

Table 3-4. Antenna Size Versus LNA Noise Temperature

Antenna Size		LNA Required (K)
8 ft	(2.5 m)	100
10 ft	(3 m)	120
12 ft	(3.7 m)	200
15 ft	(4.6 m)	300
20 ft	(6 m)	600

SUMMARY

The antenna and LNA are two of the most important components in any TVRO installation. They alone virtually determine the system's video signal-to-noise ratio, and hence determine the quality of the received picture. To improve system performance, quite possibly all that might be necessary would be an increase in the size (gain) of the antenna, or reduction in the noise temperature of the LNA.

In the following chapter, we will take an in-depth look at the TVRO receiver itself with an eye toward familiarizing you with the device from a block diagram point of view. At that point, you should be quite comfortable with the TVRO system and how it works.

Chapter 4

The Satellite TV Receiver

The first three chapters of this book have contained information about satellite communications in general, and some particulars about system components. This chapter will present the final component in that communications system—the receiver. There is no intent to endorse any of the equipment shown or discussed here for two reasons. First, the purpose of this writing is merely to provide information and education concerning receiver technology. Second, by the time this is read, some of the particular descriptions or examples given may be out of date. Such is the case with technology today. Principles are constant, but applications vary. We intend, therefore, to concentrate on those principles.

RECEIVERS

Nearly everyone has been exposed to one kind of communications receiver or another. Eyes receive and detect light waves. Ears receive and detect sound waves. There are millions of AM/FM radios, stereos, televisions, etc., all over the world. The air has been filled with radio waves and consumers have been flooded with equipment capable of receiving and detecting those signals.

What is a receiver? AM/FM stereo radios are certainly a form of communications receiver. When component stereo came along, however, this single piece of equipment was sometimes replaced with a separate tuner and amplifier, which, when combined, would provide the same function as the original receiver. This function is, of course, to accept modulated RF signals, select the station of interest, possibly convert this signal to a new intermediate frequency where filtering and amplification might occur, detect or demodulate the signal, amplify this baseband signal (sometimes audio), and supply it for use. A satellite receiving system contains these same basic functions, but they are sometimes distributed in a different fashion, much like the component stereo system. Figs. 4-1 and 4-2 show a comparison of these receiving systems. The receiver in Fig. 4-1 is contained in one box and may be located at your bedside, in your pocket, or on your wrist. The receiver shown in Fig. 4-2 is a very similar system, but the components are distributed. This system may consist of a separate antenna, a frequency converter, a "receiver" (really a detector or demodulator), and a television set. Some manufacturers choose to call their equipment by their separate functions, such as downconverter, tuner, demodulator, modulator, monitor, etc. When investigating such systems, it is important to know what elements that particular manufacturer has included in his "receiver" or "demodulator."

Let's quickly examine the typical satellite TV system as shown in Fig. 4-2. The signal is first gathered by a parabolic dish antenna as described in Chapter 3. These dish antennas are becoming common sights at hotels, motels, and even private homes. Next, the signal is amplified, frequency converted, and the desired channel is selected or "tuned-in." We now have selected one transponder from a given satellite and have converted the frequency of the RF carrier, with its information, to a new intermediate frequency or IF. The next component, which is often called the receiver (a misnomer), will remove this information in the demodulation process. This yields *baseband* audio and video. The term *baseband* is used to describe the fact that no more frequency conversions will take place. These baseband signals consist of audio frequencies from 15 Hz to 15 kHz and video information from 30 Hz to 4.2 MHz. The audio may then be amplified and fed to a speaker, while the video may be passed to a baseband video monitor. The use of a monitor for video is not too common in home applications because most users have an ordinary TV set rather than a video monitor (as component TVs become popular, this may change). A television commonly accepts signals in the 50–900-MHz range; channels 2–83. As a result, the solution is to take this audio and video baseband information and add it to an RF carrier on some standard TV channel, like channel 3 or 4. The component that does this is called the *modulator.* The signal then looks just like a conventional TV station to a television set.

The preceding discussion should give you a picture of the typical

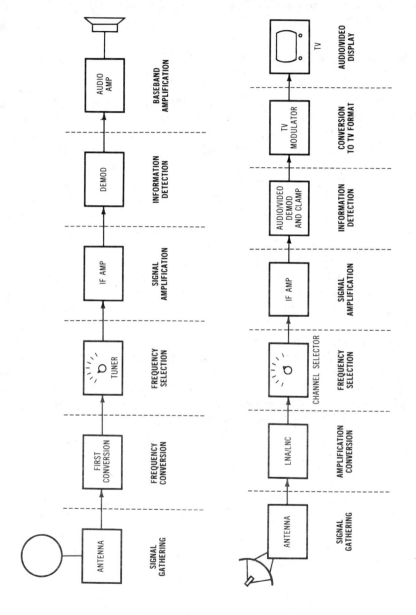

Fig. 4-1. Block diagram of an AM radio receiver.

Fig. 4-2. Block diagram of a satellite TV receiver.

TVRO system. The actual names of individual components and their placement within the system may vary among manufacturers. We will now take a more in-depth look at the receiving system.

THE DOWNLINK SIGNAL

As was described in previous chapters, the 4-GHz downlink signal from the satellites is very weak, on the order of −135 dBm (much less than the power required by a television receiver), so it must be greatly amplified. Because the signal is so weak, or "down in the noise," the first amplifier is very critical and requires a special device called a GaAs FET, or gallium-arsenide field-effect transistor. Several GaAs FETs may be used in this component, called a low-noise amplifier (LNA) or low-noise converter (LNC). These components were discussed in detail in Chapter 3. Most often the LNA or LNC is located at the antenna. In the case of the LNA, both the input signal and output signal are at 4 GHz. Because of this extremely high frequency of operation, very expensive low-loss coaxial cable and connectors must be used to carry the signal to the next component in the system, the frequency converter. This is most often a very low-loss coaxial cable such as RG-8/U, RG-214/U, heliax, etc. The type of cable that can be used depends on the length of the cable run, and costs may vary from 50 cents per foot to over $3.00 per foot. In addition, rather expensive connectors, called N-type connectors, are commonly used to minimize loss of signal.

As a rule, the higher the frequency of operation, the greater the care and expense that is required to handle the signal. In the end, the frequency of the signal must be reduced significantly to be processed (demodulated). Some have asked, "Why not convert the signal to a new lower frequency early in the string of components and eliminate some of the costly components?" Several manufacturers have undertaken this task in two different ways. One is to use a frequency-agile or tunable converter placed outside at the antenna (this unit can be placed inside or outside). If used outside, it must be protected from the elements because it is a very sensitive device. With this method, one transponder or channel is selected at a time and the signal is sent to the next component. The agile (tunable) converter method usually converts to 70 MHz, a relatively low and easy-to-handle frequency. The disadvantages are: (1) This remote agile converter must somehow receive its tuning information, usually through a multiconductor cable; (2) these rather sensitive components must be placed outside, subjected to a potentially harsh environment; (3) all "receivers" fed by this system will show the same channel (not always a concern except where multiple users desire different channels).

The other method is to convert the entire 500-MHz block of satellite TV frequencies, 3700 to 4200 MHz, down to a lower block of frequencies using an LNC. Different manufacturers have selected different frequency blocks for downconversion; common ones are 270–770 MHz, 940–1440 MHz, and 1000–1500 MHz. The idea here is to amplify the signal and lower the frequency to allow simpler handling, provide access to more than one channel simultaneously at an indoor location, and not require the tuning device to be remotely located. Thus, it overcomes the mentioned disadvantages of the remote agile converter. The 270–770-MHz version offers the least expensive system approach (cable, connectors, splitters, and amplifiers) because of the low frequency, provided the manufacturer has utilized good design techniques. The block conversion (LNC) method also offers the most flexibility and allows for multiple receivers to be tuned to different channels.

One method used by early TVRO operators was to place the LNA outside and bring the 3.7–4.2-GHz signal indoors. An agile converter was then located inside, sheltered from the elements. This method required more caution and expense (in cable and connectors) because the users were dealing with microwave frequencies. Placing the agile converter outside is merely an evolution of this method, whereas the block conversion technique is indeed a different method. Fig. 4-3 illustrates these two configurations.

These methods have definite advantages and disadvantages. Be sure to evaluate what will satisfy your particular system requirements if you are making a selection. Depending on the manufacturer and the design, each of these types of systems can perform extremely well or be a dismal failure.

CHANNEL SELECTION

In the last section, we examined an agile converter, sometimes called a channel selector, or tuner. These terms all describe the device used to select the desired channel on a satellite receiver. In this section, we will talk some more about the particulars of that device.

AM and FM radios all have tuners of one form or another. Some use a round scale with an indicator (arrow) showing the frequency selected. Others use a "slide-rule" type indicator. Televisions have typically used two types of tuners. VHF TV tuners had a "detent" dial,

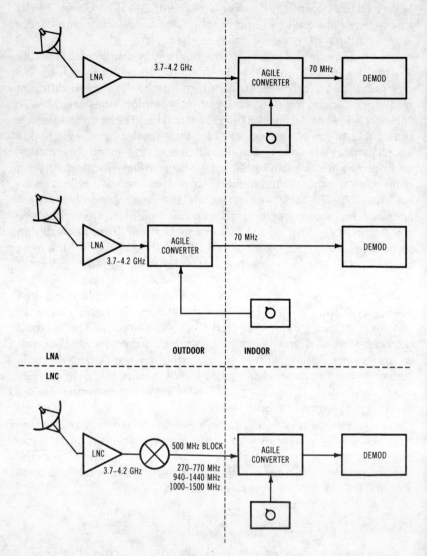

Fig. 4-3. Satellite TV system configurations.

or discrete tuning that would step from one channel to another. UHF tuners were sometimes continuously variable from channel 14 through channel 83. These tuners were constructed in particular ways, mostly because of technology and convention.

If you look at the FM radio spectrum in Fig. 4-4, you will see that the FCC has allocated station channel assignments every 200 kHz from 88.1

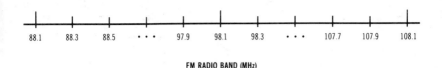

FM RADIO BAND (MHz)

Fig. 4-4. The FM radio band.

MHz up to 107.9 MHz. Historically, FM receivers have contained a slide-rule type of frequency selection which forced the user to continuously tune through unused frequencies while tuning from station to station within the FM band. Recently, however, through the use of digital technology, even FM receivers have moved toward discrete frequency selection techniques. In other words, many FM receivers today tune only to the discrete frequencies of 88.1 MHz, 88.3 MHz, 88.5 MHz, etc., up through 107.9 MHz in discrete 200-kHz steps. This allows the user to select all possible FM channels for reception. All the frequency space in between these channels is unoccupied. So why even listen or tune between channels? Well, it hasn't been until recently that technology has made "discrete" selection of frequencies economical. We are now able to tune precisely to each frequency desired and skip all frequencies in between. This technology applies to tuners of all types—AM, FM, TV, satellite, shortwave, etc. One particular type of digital channel selection employs a *phase-locked-loop frequency synthesizer*.

Entire books have been written on phase-locked loops (PLL). Their operation will not be covered here, but Fig. 4-5 is included to show a basic block diagram of a PLL synthesizer for those who desire a little more insight. A word of caution: The term PLL has become a "buzzword" in all areas of consumer electronics. PLLs are used in a variety of ways in electronics, not just in frequency selection circuits. Many sales people have no idea what a PLL is or how it is used in the product they are selling. Don't let the inclusion of the term PLL convince you of the superiority of a product. There is a lot of good equipment available without this technology, and it functions very well.

THE TUNER

Let's take a step back for a moment. What is this "tuner" we've talked about? It is almost always a variable frequency oscillator of some sort. The operator's interface is a tuning knob that controls the frequency of this oscillator. The output from the oscillator is then "mixed" with the incoming signal and the sum or difference frequency is further processed and finally demodulated to recover the desired information.

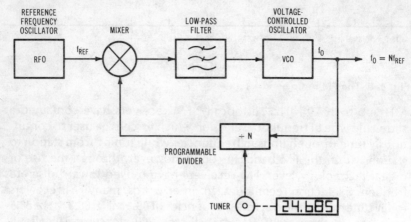

Fig. 4-5. Block diagram of a phase-locked-loop synthesizer.

Fig. 4-6 shows a simplified block diagram of a typical satellite tuner. Remember, this is a simplification for instructional purposes. In reality, it will likely be more complicated.

Just as in the FM spectrum, the FCC has assigned certain channels, or precise frequencies, for the transmission of satellite TV signals. The tuner can be designed such that the receiver will cover the entire 500-MHz satellite band from 3700 MHz to 4200 MHz, or it can be designed to select only the 24 available channels, spaced every 20 MHz from 3720 MHz to 4180 MHz.

Now we will examine some different ways of implementing the oscillator. Phase-locked loops are just one type of tuning method used in satellite TV channel selectors. They are likely to be the most stable or drift-free due to their frequency and phase-locked design. The accuracy, however, is dependent on the overall stability of an internal reference oscillator. Another common type is the "frequency-preset" type. Push-button car radios are a mechanical version of this type of frequency selection. Some popular televisions and VCRs have a finite number of "preset channel" slots available. These are separate tuners or tuning elements, each preset to the desired channel. After presetting these the user electronically or mechanically steps from one to another. Another preset type uses fixed or variable resistors to tune an oscillator. There is, of course, the old standby—a continuously variable type tuner.

All of these mentioned are subject to drift, or slight changes in frequency. By the very nature of such devices, they are made to be variable at the user's discretion. An almost unavoidable side effect is that they may also change frequency on their own due to mechanical

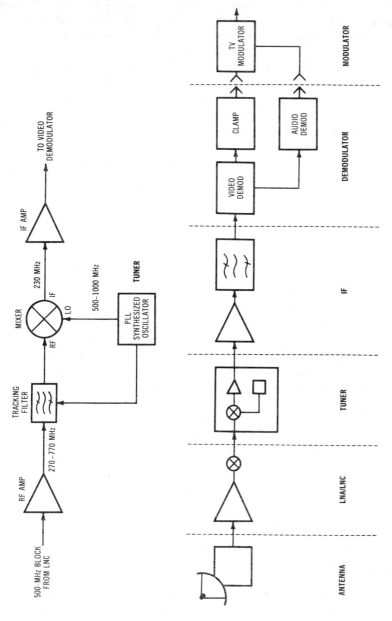

Fig. 4-6. Block diagram of a typical satellite tuner.

Fig. 4-7. More detailed block diagram of a satellite TV receiving system.

113

assembly, electrical properties, and Mother Nature. That means that as the tuner heats up, cools off, or ages, the oscillator may change in frequency and gradually drift away from the originally selected station. This is often compensated for by a "fine-tune" control. The drift of the oscillator can be tolerable if it is small, but becomes quite aggravating when it must be constantly retuned. One way to partially compensate for this drift is called AFC (automatic frequency control). This has been used in FM radios for a number of years and can be utilized with certain effectiveness in satellite TV receivers.

Some of the tuners mentioned can be remotely operated. There are two basic types of "remote controls." The least complicated (sophisticated) is a "hard-wire remote." That means that a group of wires is strung between the operator's control box and the actual tuner. Sometimes this can be awkward as the wires may get in the way. Another version of this remote control is the carrier-current remote which converts the tuning information to a signal that is passed over the house's electrical wiring circuits. In this case, the user must locate himself and the controller near an AC outlet. The most popular method, and that used in most remotely operated televisions, is the wireless remote control. This sends the tuning information out over an infrared, sonic, or RF transmitter.

For most home satellite TVRO systems, any of the previously mentioned tuning schemes will be adequate if properly designed and built. Under some input signal conditions, however, they will not be adequate. An example is in the transmission of digital data over the satellite, as in the case of Reuters (the news service) or SNC (Satellite News Channel). Here, frequency stability is very critical. For most simple video systems, however, stability won't be a problem. As has been mentioned over and over in this book, you get what you pay for. If hands off, trouble-free operation is desired, it costs.

THE IF

Because Fig. 4-2 is an oversimplification of a satellite receiver, Fig. 4-7 is included to give a more detailed view. Most radios of any sophistication have an IF or intermediate frequency where processing of the signal takes place.

An RF signal at the input of a receiver is frequency converted, or mixed, to a different frequency called the IF. This is done primarily to convert the signal to a frequency or range of frequencies where the signal can be more easily processed. A common IF in AM broadcast radios is 455 kHz, in FM broadcast radios it is 10.7 MHz, and in many satellite TV receivers it is 70 MHz. There may be more than one IF in a radio, and the frequencies may vary depending on the processing that is to be performed. IFs may also be found in both transmitting and receiving equipment.

Two types of processing typically take place in the IF—amplification and filtering. We will briefly discuss these two in the following paragraphs.

Amplification

Consider the strength of the signal that is present in a satellite system. With the +65-dBm EIRP from the satellite's antenna, and the 200 dB of path loss in the downlink propagation path, the signal level present at an earth station antenna is approximately −135 dBm. The combination of dish antenna and LNA/LNC can boost this level by approximately 100 dB to −35 dBm. From 10 dB to 30 dB may be lost in the frequency conversion from RF to IF. Additional losses resulting from filtering can vary from 10 to 35 dB or more. Thus, the combined losses can then vary from 20 dB to 65 dB depending on the techniques employed, resulting in a signal level (with no amplification) of −55 to −100 dBm at the input to the demodulator. The demodulator, however, will most likely require a signal level on the order of 0 dBm. That means the IF will require from 55 to 100 dB of gain, which will be spread out over several stages of amplification throughout the frequency converter, IF, and demodulator circuits.

Filtering

Another critical operation performed in the IF is filtering. As a result of the mixing process, in addition to the desired signal there may be many undesired signals. Filters can be designed to remove all but the desired signal. In doing this, great care must be taken not to alter the characteristics (bandwidth, phase, linearity, etc.) of the signal of interest.

Although the bandwidth of a satellite transponder may be 40 MHz, the bandwidth of the filter used in a receiver will vary from 17 MHz to 32 MHz. These filters pass both signal and noise, two components that determine a performance characteristic called the *signal-to-noise ratio* (S/N or SNR). The object is to narrow the filter bandwidth and pass less noise, thereby increasing the ratio of signal to noise. This works until the filter begins removing information contained in the signal (video and all subcarriers) as well as removing the noise. In other words, the bandwidth of the IF filter must be optimized for the amount of information contained in the desired signal. The filters employed in the IF will determine many of the overall performance characteristics of a satellite receiver and thus are very critical circuit elements.

In summary, the IF is that frequency at which the bulk of amplification and filtering is accomplished. IFs are found in all types of communications equipment from radar to AM radio. The frequency or band of frequencies chosen as the IF is selected according to physical and electrical constraints identified by the designer.

DEMODULATION

The antenna and LNA or LNC are discussed in depth in Chapter 3. We have just talked about the tuner and the IF and their places in a receiving system. The next logical topic in a receiver discussion is the demodulator. That is the device which recovers *baseband* information from the radio frequencies used as carriers.

The television signals that are broadcast directly into homes via local TV stations have both video and audio information which must be recovered. The video is added using amplitude modulation (AM), and the audio is added using frequency modulation (FM). Television signals broadcast over the satellite, however, have all audio and video information added by FM techniques. That is one of the reasons that a dish antenna cannot be attached to the back of your television for satellite TV reception. In addition, the satellite signal carries much more information than a typical TV signal. Several audio subcarriers may be included for multiple language broadcasts, stereo signals, or news broadcasts.

The intermediate frequency output from the converter is fed through the IF and then directly to the demodulator. Fig. 4-8 shows the frequency spectrum of this baseband signal. The video information is contained from 30 Hz to 4.2 MHz. This is standard in the United States and will seldom deviate.

The FM video can be suitably demodulated in a number of different ways. The most common are either via an FM discriminator, or a phase-locked-loop demodulator. Selection of these is up to the designer and system constraints. The frequency of the incoming signal and the performance characteristics required will have a lot to do with the circuit employed. The demodulated video out of the FM detector is actually called *composite baseband video*, because it is composed of baseband video information and one or more audio subcarriers. This signal is passed on to the video clamp circuit and to the audio demodulator circuit for further processing. First, we will discuss the clamp.

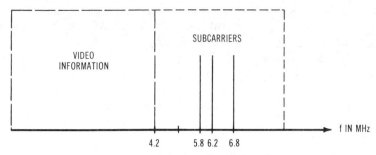

Fig. 4-8. Frequency spectrum of the baseband signal.

According to certain broadcast standards (CCIR), the video signal transmitted to the satellite has its higher-frequency components emphasized (preemphasis) in order to compensate for a phenomenon encountered in FM demodulation in which the noise output of the demodulator is greater at the higher video frequencies. To balance or reduce this noise, a low-pass filter, called a *deemphasis filter*, is placed in the circuit. This lowers the level of the higher video frequencies, which results in an almost flat noise response from the FM demodulator. It also removes much of the audio subcarrier signal. A look at the video at this point would yield something like Fig. 4-9. This shows the video signal riding on a triangular waveform called the *energy dispersal waveform*. As was discussed in Chapter 2, this technique is used to help prevent interference to terrestrial microwave communications systems. For example, if the video information on the RF carrier being transmitted over the satellite were ever lost, the satellite transponder's carrier would concentrate all of its energy at one frequency instead of spreading it out over the transponder's bandwidth. Because there are terrestrial microwave systems operating in the same frequency band, it was feared that these high-powered, unmodulated carriers could cause severe interference. In order to avoid this potential problem, the dispersal waveform was added. It is simply a 30-Hz triangular wave added to the video at the broadcast transmitter. The receiver must "clamp," or remove this triangle wave in order to produce a suitable video signal. The clamp is a relatively simple circuit; however, its operation sometimes separates the good from the marginal receivers. If it is inadequate, the picture will flicker at a 30-Hz rate.

The entire demodulation process isn't complete, though, until we recover the audio information. This information may be contained on multiple FM subcarriers ranging from 5 to 8 MHz. As a result, most manufacturers provide audio demodulators for one or more of these subcarriers as standard in their receivers. The most common are on 6.2

and 6.8 MHz. Recently, many broadcasters have begun to put information on other subcarriers in addition to these two. As a result, it is highly desirable to have a tunable audio demodulator. This way a single audio demodulator can select from a number of subcarriers within the 5- to 8-MHz range. It may also be desirable to have more than one audio demodulator, in order to simultaneously receive both English and French broadcasts on the Canadian ANIK satellites. Reception of some stereo music broadcasts will require two audio demodulators, one for each channel of stereo information.

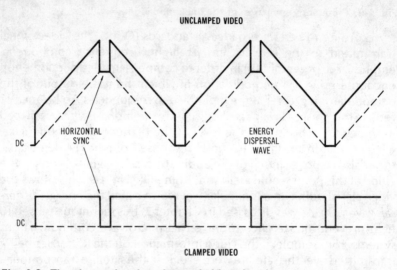

Fig. 4-9. The clamped and unclamped video signal.

Each audio demodulator uses a filter to select only one of the audio subcarriers in the 5–8-MHz range. This subcarrier is then demodulated and is made available as a 15-Hz to 15-kHz audio output from the receiver.

THE OUTPUT SIGNALS

The last section discussed the demodulation process or removal of information from the RF carrier. There are essentially three signal outputs. The output from the video demodulator is called *composite baseband*, consisting of both audio and video information. When this signal is then processed in the clamp, only *baseband video* remains. This same signal passed through an audio or subcarrier demodulator yields *baseband audio* information. At this writing, however, the majority of television receivers accept only VHF and UHF input signals.

So the video and audio baseband information must be converted to a standard TV signal, usually at channel 3 or 4. The device that does this is called a *TV modulator.* They are very common devices and can be purchased at many electronics outlets. In fact, home computers often require these same modulators to drive television sets. Most satellite TV receivers do not include the modulator circuit as a standard feature. If you are fortunate enough to have a video cassette recorder, however, it will most likely accept these audio and video outputs and provide a modulated RF output on channel 3 or channel 4. Also note that as component television becomes popular, each monitor may accept the audio and video directly without a TV modulator. Otherwise, it will most likely be necessary to purchase your own modulator.

It is difficult to predict what will happen in the future with this or any other high technology industry. With the growth in the home satellite TVRO industry, the pay channels such as HBO, Cinemax, Showtime, etc., are losing revenues. There is serious talk of these programmers altering their signal through scrambling so that most private, nonpaying viewers can no longer watch. In this event they will likely sell or lease equipment capable of descrambling their signals. In order to do this, the composite baseband signal output from the video demodulator may have to be accessible as an output from the home receiver. It is this signal that will most likely be applied to a descrambler. Keep this in mind while reviewing features of satellite receivers.

SUMMARY

In this chapter, we have tried to present a very technical subject in a less than technically rigorous manner. A satellite receiver can be viewed in a rather simple form. It accepts an FM signal from either an LNA (3.7–4.2 GHz) or LNC (500-MHz UHF block), selects one channel, amplifies and filters it, demodulates or removes the audio and video information, and provides these as outputs. These can then be displayed on a conventional television set. That was fairly painless. What has been left out are all the engineering design details. If you still have questions, consult the manufacturers or any of the available texts and periodicals on this subject. There are many manufacturers of satellite television receivers. They provide equipment whose performance will vary from marginal to absolutely flawless. To a very great degree, this will be reflected by price. Eventually, the marginal equipment will vanish and only those acceptable by the majority will survive. At that time, prices will stabilize and risks will diminish. Until that time, carefully review the products before making a decision.

Appendix A

Glossary

amplifier—A device used to increase the level of an electronic signal.

amplitude modulation (AM)—A modulation technique in which the amplitude of the RF carrier is made proportional to the level of the information signal.

antenna—A conducting surface used to transmit or receive electronic signals.

attenuator—A device or component in a system that causes a reduction in the level of a signal as the signal passes through it.

audio subcarrier—An RF carrier, typically between 5 and 8 MHz, which is modulated with audio and subsequently combined with the video carrier and transmitted over the satellite.

automatic frequency control (AFC)—A circuit that locks the receiver to the chosen frequency and prevents the receiver from drifting away from its assigned channel.

automatic gain control (AGC)—A circuit used to control the gain of an amplifier in order to maintain a constant output signal level when the input signal level of the amplifier fluctuates.

AZ-EL mount—One type of mount for an antenna system that requires two distinct adjustments (azimuth and elevation) to move the aim of the antenna from one satellite to another.

azimuth—A magnetic compass direction, expressed in degrees, after correction for magnetic deviation.

bandpass filter—A device or circuit that passes certain groups of frequencies while blocking or greatly attenuating all others.

bandwidth—The range of frequencies that is allowed to pass through a device or circuit with minimal attenuation, usually 3 dB.

baseband—The signal used for actual informational purposes and not riding on an RF carrier. Video baseband contains signals in the range from 30 Hz to 4.2 MHz. Audio baseband is from 30 Hz to 15 kHz.

beamwidth—The angle or conical shape of the beam the antenna projects. Larger antennas typically have narrower beamwidths and are, therefore, more difficult to aim at a given satellite.

block downconversion—The process of converting a wide band of input signals to a lower intermediate frequency. For example, conversion of signals in the 3.7- to 4.2-GHz band down to a lower frequency between 270 and 770 MHz.

carrier—A high-frequency radio signal that can be modulated with baseband information for transmission through a conducting medium.

carrier-to-noise ratio (C/N)—The ratio of the RF carrier level to the noise level of a signal in a given bandwidth, usually measured in dB.

Cassegrain—An antenna comprised of two reflectors; the parabolic reflector (dish) and a hyperbolic subreflector at the focal point of the dish. The hyperbolic subreflector reflects signals back into the feed.

CATV—An abbreviation for cable television. It was formerly called *Community Antenna TV.*

C band—The band of frequencies in the 3.7- to 4.2-GHz range.

chrominance—The property of light that produces the sensation of color in the human eye.

Clarke belt—The orbital locations around the earth in which all geostationary satellites reside.

coaxial cable—A transmission line comprised of a center conductor surrounded by insulation and then wrapped with a conductive braid, all of which is then sheathed in another outer jacket of insulation.

compass heading—The magnetic direction as read on a compass, expressed in degrees.

composite video—For color signals, this consists of blanking, field, and line synchronizing signals, and luminance and chrominance picture information all combined into a single signal.

contrast—The ratio between the maximum and minimum brightness levels of a picture.

cross polarization—The condition that exists when the polarization of an antenna/LNA/LNC and the polarization of the signal it is trying to receive are not perfectly matched. There is always some degree of cross polarization within a system. It should be minimal, however.

dBm—A logarithmic relationship comparing some power level (P1) to 1 milliwatt: dBm = 10 log (P1/1 mW).

DBS—An abbreviation for Direct Broadcast System. This is a proposed low-cost, K_u band (12 GHz), over-the-satellite broadcast system in which each affected home would have a small roof-mounted, one-meter dish antenna, and a low-cost converter to convert the signals to a format compatible with a normal TV.

dBW—A logarithmic relationship comparing some power level to one watt: dBW = 10 log(P1/1 W).

decibel (dB)—A logarithmic ratio used as a type of shorthand notation to express very large or very small numbers.

declination—An angular measurement in degrees relating to a deviation error in polar-mounted antennas.

deemphasis—The modification of a signal by amplifying one range of frequencies with respect to another in order to remove any pre-emphasis.

demodulate—The process of recovering a baseband signal from a carrier.

directional coupler—A device used to pass a signal in one direction with almost no attenuation, while simultaneously passing that signal to an isolated port at some specified value of attenuation.

dish—The parabolic reflecting system of an antenna.

dish illumination—That portion of a dish that is actually seen by the feed. The feed should not be capable of seeing beyond the outside edge of the dish since such operation will produce excess system noise.

dispersal waveform—A 30-Hz triangular waveform that is added to the uplink video signal to disperse or spread the RF spectrum of the transmitted signal. It helps to prevent interference both to terrestrial microwave systems, and between the signals passing through the satellite.

dither—An effect seen as flicker on the screen of a television at a 30-Hz rate, which may be caused by an insufficient removal of the 30-Hz dispersal waveform by the clamp in the receiver.

downconverter—A device used to convert high-frequency signals to lower-frequency signals. This conversion may be from 4 GHz to something much lower, such as 70 MHz, 270 MHz, or 1 GHz.

downlink—The path from the satellite transmitter and transmitting antenna to the earth station receiving antenna. The frequency of operation for the downlink propagation path is from 3.7 to 4.2 GHz.

dual conversion—To convert a signal at one frequency or group of frequencies to another frequency or group of frequencies in a two-step process. The resulting frequency after the first conversion is called an *intermediate frequency*.

dual orthomode coupler (or transducer)—Provides both horizontal and vertical polarization on a receiving antenna in order to drive two orthogonally polarized LNAs.

earth station—A transmitting or receiving station located anywhere on this planet to be used for satellite communications.

EIRP—An abbreviation for effective isotropic radiated power, the power that is effectively radiated from an antenna given the power into the antenna and the gain of the antenna.

elevation—The angle above the horizon that a transmitting or receiving antenna must be aimed in order to communicate through a satellite.

F/D ratio—The ratio of the focal length of an antenna to the diameter of the antenna. Given a fixed diameter, the larger the F/D ratio is, the shallower the dish must be.

feed—The device mounted at the focal point of the antenna which collects the signals reflected from the dish and "feeds" them to the LNA.

field—One-half of a complete picture (frame) of a displayed television signal which contains all of the odd or all of the even scanning lines.

focal length—The distance, as measured directly in front of the dish, from the center of the dish to the point at which all reflected energy is focused.

focal point—The place at which all reflected energy from a dish is directed.

footprint—The geographical area on the earth's surface that is covered by signals radiated from a satellite.

frame—One complete television picture consisting of two interlaced fields.

frequency modulation (FM)—A system of modulation in which the instantaneous frequency of an RF carrier varies in proportion to the amplitude of a modulating signal.

frequency reuse—A technique for reusing a given band of frequencies by transmitting independent information on orthogonal polarizations.

gain—The amount of amplification that a particular device may have.

geodetic south (true south)—Magnetic south corrected for magnetic deviation.

geostationary—Stationary with respect to a fixed reference on the surface of the earth.

geosynchronous—An orbit that is synchronized with the rotational rate of the earth.

G/T—A figure of merit given to receiving earth station systems. It is the ratio of the system gain to its noise temperature in dB per kelvin. The higher this ratio is, the better the system performance will be.

hardline—A type of very-low-loss coaxial cable made with a full metal shield rather than the conductive braid found in flexible coaxial cable.

headend—The central point of a cable TV system where VHF, UHF, FM, and satellite signals are processed, combined, and fed into the system.

HF—An abbreviation for high frequency. It usually describes those frequencies in the range from 3 to 30 MHz.

HPA—An abbreviation for high-power amplifier. It is found as the final stage in most transmitting installations.

image—The result of the mixing process. When an RF signal is mixed with an LO signal, both the sum and difference frequencies result. One is selected as the IF, the other is the image.

intermediate frequency (IF)—A new frequency to which a signal is shifted as an intermediate step in the overall modulation or demodulation process. It is this frequency at which most of the signal processing takes place.

isolator—A device that allows signals to pass in one direction but not in the opposite direction.

kelvin—A unit of measure for absolute temperature. Zero kelvin is absolute zero, meaning it is the temperature at which all molecular motion stops. There is no colder temperature than zero kelvin.

K$_u$ band—The satellite frequency band in the range 11.7 to 12.2 GHz.

latitude—A distance, measured in degrees and minutes, north or south of the equator.

local oscillator—A frequency source used to mix with an RF signal to produce the IF.

longitude—A location measured in degrees and minutes east or west of the Greenwich Meridian in England.

low-noise amplifier—A device used to amplify the very-low-level signals that are focused on the feed by the dish antenna. Its primary purpose is to amplify the signal without adding any additional noise to the system.

low-noise converter—A combination LNA and downconverter in the same package.

LPTV—An abbreviation for low-power television.

luminance—The amount of light intensity in a given picture perceived as brightness to the human eye.

magnetic deviation—An angular adjustment in degrees between a compass indication and the true or geodetic direction.

MATV—An abbreviation for master antenna television. An antenna installation used by a local community to receive television signals from various sources, combine them, and distribute those signals to the community through a cable network.

MDS—An abbreviation for multipoint distribution system. A microwave (2 GHz) TV distribution system that uses an omnidirectional transmitting antenna to supply TV signals throughout a community. It requires a low-cost antenna/converter system in front of your TV in order to receive the signals.

microwave—A term used to describe an RF signal that has a very short wavelength, less than 10 cm, corresponding to frequencies between 3 GHz and 30 GHz.

mixer—A circuit whose two inputs (RF and LO) are "mixed" to produce sum and difference frequencies, one of which is chosen as the IF or intermediate frequency.

modulate—To place information on a carrier for long-distance transmission. This is done by varying a certain characteristic (amplitude, phase, frequency, etc.) of the carrier with changes in the baseband information.

modulator—In a receiving earth station, a device used to convert the baseband video and audio information out of the receiver to a new frequency that can be received by a standard TV set.

noise figure—A figure of merit for an amplifier which indicates how much noise it adds to a signal that passes through the device. The lower the number is, the better is the amplifier.

noise temperature—Another method of expressing the amount of noise that a device will add to a signal as it passes through the device. Usually expressed in kelvins.

orthomode transducer—A device attached to an antenna feed which allows the antenna to be used simultaneously for transmission and reception. The signals must be on orthogonal polarizations, however, because the transducer is polarization sensitive.

path loss—The loss incurred when a signal travels from one point to another through space.

polar mount—An antenna mounting mechanism which allows the antenna to be swept over all or a portion of the satellites in the Clarke belt with one motion (as opposed to the two movements required of an AZ-EL mount).

polarization—The orientation of the electric field of a traveling wavefront. For satellite TV signals, the orientation is usually either horizontal or vertical.

polarization efficiency—A measure of the efficiency of an antenna due to the misalignment of the antenna's polarization with respect to the polarization of the signal that is to be received.

preemphasis—The modification of a signal by emphasizing one range of frequencies with respect to another. It is usually used to improve the signal-to-noise ratio in FM systems.

prime focus—A type of antenna that has a single reflector (dish) which tends to focus all radio energy incident upon the dish to a single point. The feed is then located at this point.

side lobes—A portion of a beam from an antenna other than the main lobe. For a receiving antenna, these lobes tend to pick up extraneous signals from directions other than where the antenna is aimed.

signal-to-noise ratio—The ratio of the baseband signal level to noise level in a system. It is a figure of merit for signal degradation due to noise.

site survey—An evaluation of a proposed site for a receiving antenna. The survey will verify whether a satellite can be "seen" from your proposed location, and will also predict the amount of terrestrial interference your system might encounter.

SMATV—An abbreviation for satellite master antenna television. It is a minicable TV system typically serving hotels, motels, apartments, and condominiums, the heart of which is a satellite TVRO earth station.

space loss—Otherwise known as spreading loss, it is the loss that occurs to every transmitted signal due to the spreading of its wavefront as the transmitted signal travels further and further from the point of origin.

splitter—A device which divides an input signal equally between two or more outputs.

spherical antenna—A dish-type antenna where the curvature of the dish resembles that of a sphere as opposed to a parabola.

terrestrial interference—Interference to satellite TVRO stations caused by high-power land-based microwave links in the 4-GHz band. The Bell system is often the culprit.

threshold—The level of RF carrier-to-noise ratio required to produce an acceptable S/N ratio in the received signal.

threshold extension—A system that lowers the threshold of an FM demodulator to allow it to operate with a lower C/N ratio without noise showing up in the picture.

transponder—A device located on the satellite that is used to receive a communications signal from earth, translate the signal to a new frequency, and transmit the signal back to earth. A receiver and transmitter back-to-back.

trap—A band-stop filter used to attenuate all signals within its trapping range.

TVRO—An abbreviation for television receive only. Usually refers to an earth station receiving installation.

UHF—An abbreviation for ultrahigh frequency. The frequency range from 300 to 3000 MHz.

uplink—The propagation path from the earth station transmitting facility to the satellite on the first leg of the signal's journey.

vestigial sideband—A modulation scheme in which almost all of one sideband of an amplitude-modulated signal is removed. The fractional sideband that remains is only a "vestige" of what was once there, hence the terminology.

VHF—An abbreviation for very-high frequency. The frequency range from 30 to 300 MHz.

video monitor—A device, similar to a television set, which displays television pictures but which can accept only baseband video and audio rather than a modulated carrier as its input.

VSWR—An abbreviation for voltage standing-wave ratio. It is a measure of the quality of an impedance match between two devices.

waveguide—A very precise hollow metal tube used to guide very-high-frequency signals from point to point. It is used at microwave frequencies because of its low loss when compared to coaxial cable.

Appendix B

Direct Broadcast Systems

The present satellite TV system in use in the United States operates in the 4- to 6-GHz range, also known as the C band. This service has been intended primarily for transmission to broadcast and cable TV (CATV) installations. Home TVRO installations were not originally planned as a part of this C-band system, although many thousands of these "pirate" installations are in operation today. In fact, it has long been thought that the C band would not be suitable for direct-to-the-home broadcast because of the large size of the antenna required for acceptable picture reception (see Chapter 3). The feeling was that an antenna of this size would be both too expensive and too esthetically displeasing to place in the average consumer's backyard.

In order to eliminate the need for a relatively large dish antenna, and to reduce the potential cost to the consumer, direct broadcast into the home via satellite at an operating frequency of 12 to 14 GHz (K_u band) has been proposed. Antennas for home TV reception at this frequency would be on the order of 2 to 3 feet in diameter instead of the 6 to 10 feet required at the C band.

Satellite Business Systems (SBS) of McLean, Virginia already has a satellite in orbit which operates at the K_u band, but it carries very little television programming. Instead, its primary function is the transmission of high-speed data for use in business applications (banking, insurance, etc.). Canada and Japan, however, are using the 12- to 14-GHz band for limited DBS activity. SBS and Telesat of Canada, for example, recently launched two K_u band satellites on board the NASA space shuttle, Columbia, in November 1982. The Canadian satellite, ANIK-C3, will be used to broadcast K_u band satellite TV signals to the vast areas of Canada. These signals will also be available across the northern United States. Imagine a 2-foot diameter dish antenna on every rooftop! Such installations are being suggested to retail in the $500 range and below.

There still remain some details to be worked out for such a multimillion-dollar system. If such a system is ever to exist, investors and manufacturers will have to provide the funds. But such a venture is risky for two reasons. First, there is currently very little consumer-oriented 12-GHz satellite receiving equipment available. The potential manufacturers of this equipment would have to invest heavily to develop these receivers not knowing how viable the system will be. Second, there is currently very little television programming available for such a system. Those who provide programming are reluctant to invest heavily in the system because there aren't sufficient numbers of 12-GHz home receivers to support their investment. (Which came first, the chicken or the egg?)

Many will argue that as technology advances, the 4-GHz antennas will decrease in size and by virtue of the number of TVRO installations and the variety of programming already in place at 4 GHz, 12-GHz DBS will never come about. The current 4-GHz system may indeed be the DBS system of the future. On the other hand, with the launch of the ANIK-C satellites, the 12-GHz system may acquire the programming it needs to support DBS. It is difficult to predict at this writing. Only time will tell.

THE DBS RECEIVER

Fig. B-1 is a block diagram for a proposed K_u band receiving system. Note that the uplink and downlink propagation paths will be similar to those of the 4-GHz systems described in previous chapters, and thus will not be discussed here. The receiving system consists of two primary units—an indoor unit (the TV set) and an outdoor unit. In this system, the outdoor unit would consist of a 2-foot diameter dish antenna, a low-noise amplifier/converter that would be used to translate the signal down to a particular channel in the conventional UHF television broadcast band, and an FM/AM converter that would convert the FM video of the downlink signal into a conventional TV format. This signal

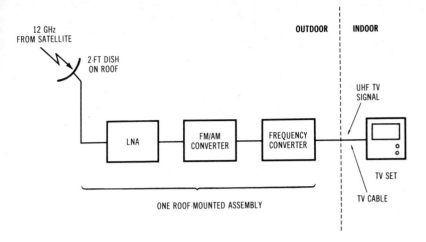

Fig. B-1. Block diagram for proposed K$_u$-band DBS receiving system.

would then be transported into the home via coaxial cable and connected directly to the television set.

Note that the outdoor unit, as presently envisioned, will be sufficiently compact and lightweight so that it can be mounted on most any rooftop with minimal effort.

131

Appendix C

Locating the Satellites and Aiming the Antenna

This appendix will attempt to explain some of the terminology and a few of the techniques in aiming a satellite TVRO antenna.

Years ago, the British established two systems of angular measurement called latitude and longitude to assist in navigation and map making. They did this by dividing the globe into 360° of longitude, starting at the Greenwich observatory in England. Traveling westward from the Greenwich meridian (0° longitude) the longitudinal axes increase from 0° to 180° of "west longitude." Proceeding in an easterly direction from Greenwich, they increase from 0° to 180° of "east longitude." Similarly, the earth was divided north and south of the equator with parallel lines called degrees of latitude, with 0° representing the position of the equator. Thus, the earth was divided into a grid-like system and any location on earth could be described by its latitude and longitude.

Prior to trying to aim an antenna at a satellite, it is imperative that you find your precise geographic location in latitude and longitude. These coordinates are available on U.S. geological survey maps, or they can be obtained from surveyors. Once you know your geographic location, it's really a very simple matter to aim the antenna.

We've already discussed the fact that all of the stationary (geosynchronous) satellites are located within the Clarke belt at a distance of 22,300 miles above the equator. Each satellite is fixed at a precise location along that equatorial belt. Its position is given in degrees of longitude. Some of these locations are listed in Table C-1.

Table C-1. Satellite Locations

Satellite	Location (West Longitude)
SATCOM 4	83°
COMSTAR 3	86.9°
WESTAR 3	91°
COMSTAR 1/2	95°
WESTAR 1	99°
WESTAR 4	99°
ANIK 1	103.9°
ANIK B	108.9°
ANIK 2/3	113.9°
ANIK C3	117.5°
SATCOM 2	118.9°
WESTAR 2	123.4°
COMSTAR 4	127°
SATCOM 3	132°
SATCOM 1	134.9°

In order to find a satellite, one of the first things that must be accomplished is to locate due south (true south). This can typically be done by first locating magnetic north with a compass and then correcting for the magnetic deviation in your area. Note that magnetic north and true north may differ by as much as 23° in some locations. Once you have found true north, rotate 180° and you will be facing due south. This is the direction that the antenna must face (in the Northern Hemisphere).

For a polar mount (Fig. C-1), the next task is to determine the angle of declination, or the elevation angle of the antenna. This can be determined from the following formula:

$$\angle = 90° - \tan^{-1} \left[\frac{3974 \sin L}{22,300 + 3974 (1 - \cos L)} \right]$$

where L is your latitudinal position in degrees, and 3974 is the equatorial radius of the earth in miles. Once the angle of the declination is calculated, the antenna's elevation angle can be found by using a device called an *inclinometer*. This device can often be found in many hardware stores. An alternative is to use a weighted string and a protractor (see Fig. C-2). In either case, the angle of elevation for the antenna is then varied until the correct angle of declination is found. Once the antenna is positioned such that it is facing due south, and once the angle of declination has been applied, the antenna should easily sweep the arc of satellites located on the equator.

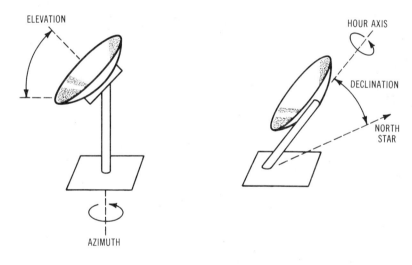

(A) AZ-EL mount.   (B) Polar mount.

Fig. C-1. AZ-EL and polar mounts.

For an AZ-EL system (Fig. C-1), the following equations can be used to aim the antenna:

$$AZ = 180 + \arctan (\tan B / \sin A)$$

135

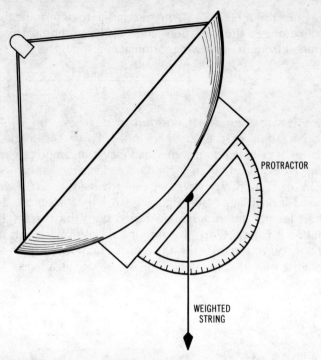

PROTRACTOR

WEIGHTED
STRING

Fig. C-2. Determination of elevation angle.

for the Northern Hemisphere, or

$$AZ = \arctan(\tan B / \sin A)$$

for the Southern Hemisphere, where A is the latitude of the earth station (north is plus, south is minus) and B is the longitude of the earth station (east is plus, west is minus) minus the longitude of the satellite.

For elevation,

$$EL = \arccos \left[\frac{D^2 + R^2 - (R - H)^2}{2 \times D \times R} \right] - 90°$$

where D is the distance between the earth station and the satellite and is equal to:

$$D = \sqrt{R^2 + (R + H)^2 - 2R(R + H) \cos C}$$

136

where,

R is the radius of the earth (3974 miles),

H is the height of the satellite above the earth (22,300 miles),

C is the central angle = arccos (cos A × cos B).

Bibliography

Barker, Rick, and Freisz, John. "Earth Station Antennas RF Considerations." *Satellite Communications Symposium,* Scientific Atlanta, 1981.

Beakley, Dr. Guy W. "Satellite Communications into the Eighties." *Satellite Communications Symposium,* Scientific Atlanta, 1981.

Chaddick, Steve. "Video Exciter Model 7550." *Satellite Communications Symposium,* Scientific Atlanta, 1981.

Cooper, Robert B., Jr. "Home Reception via Satellite." *Radio Electronics,* August 1979, September 1979, and October 1979.

Daniel, Tim. "The Satellite TV Challenge." *73 Magazine,* November 1981.

Easton, Anthony. *The Home Satellite TV Book.* New York: Playboy Press, 1981.

Gibson, Stephen. "TVRO Receivers: The Inside Story." *73 Magazine,* December 1981.

Hollis, J. Searcy. "Introduction to Satellite Communications." *Satellite Communications Symposium,* Scientific Atlanta, 1981.

Ingram, Dave. "Satellite TV Receivers." *73 Magazine,* November 1981.

Johnson, Ken. "Low Noise Amplifiers." *Satellite Communications Symposium,* Scientific Atlanta, 1981.

Kearney, Tim. "Satellite Communications." *Notre Dame Technical Review,* Spring 1974.

Rovira, Luis. "Low Noise Converter, Series 360." *Satellite Communications Symposium,* Scientific Atlanta, 1981.

Index

A

AFC, 114
AM band, 13
AM DSB versus VSB singals, 39
Amplification, 115
Amplifier, 94, 121
Amplitude modulation, 121
Anatomy
  of a dish, 63-65
  of an LNA, 92-96
Antenna(s), 121
  beamwidth, 61
  diameter versus
    beamwidth, 63
    gain, 60
  directivity, 61-63
  gain versus angle, 64
  LNA trade-off, 101-103
  mount, 80-85
  noise temperature, 68
  size versus LNA
    noise temperature, 103
  summary, 79-80
Atmospheric
  absorption, 14
  attenuation, 46
Attenuator, 121
Audio subcarrier, 44, 121
AZ-EL mount, 80-82, 121
Azimuth, 80, 121

B

Bandpass filter, 121
Bandwidth, 122
Baseband, 106, 122
  audio, 118
  video, 118
Beamwidth, 122

Block conversion—the
  low-noise converter, 100-101
Block diagram
  AM radio receiver, 107
  frequency converter, 99
  low-noise amplifier, 92
  satellite
    transponder, 51
    TV system, 30
Block downconversion, 122

C

Carrier, 122
Carrier-to-noise ratio, 122
Cassegrain antenna, 69-73, 122
CATV, 122
C band, 40, 122
Channel selection, 109-111
Clamped and unclamped video signal,
  118
Clarke belt, 22, 122
Clarke's artificial
  relay system, 17-19
Coaxial cable, 108, 122
Compass heading, 122
Composite
  baseband, 118
    video, 116
  video, 122
CONUS, 28
Converter, 98-101
Cross polarization, 122
Cryogenically cooled amplifiers, 95

D

DBS, 28, 122
Declination, 123
Deemphasis, 123
  filter, 117
Demodulate, 123

Demodulation, 116-118
Directive elements, antenna, 63
Dish, 123
    illumination, 123
Dispersal waveform, 123
Dither, 123
Downconverter, 123
Downlink, 35, 56-57, 123
    makeup, 35
    signal, 108-109
Driven elements, antenna, 63
Dual conversion, 123

E

Earth station, 124
    requirements, 22-25
ECHO I, 20
Effective isotropic radiated
    power (EIRP), 45, 124
Elevation, 80, 124
Energy-dispersal waveform, 44, 117
Explorer I, 20

F

Feed, 85-86, 124
    rotators, 86-88
Filtering, 115-116
FM threshold, 12
Focal
    length, 124
        of parabola, 64
    point, 124
Footprint, 124
    of SATCOM-IIIR, 55
Frequencies and modulation, 13
Frequency
    converter, 98
    modulation, 124
    reuse concept, 35, 52, 124
    spectrum of the baseband signal, 117

G

Gain versus smoothness, 61
Gallium-arsenide
    field-effect trasnsistor, 94
Geodetic south, 124
Geometry of
    prime-focus antenna, 65
    spherical antenna, 75
Geostationary, 124
Geosynchronous, 124

H

Headend, 125
Hertz, unit of frequency, 13
Home earth station, 26

I

IF, 114-116
Intermediate frequency, 98
Ionosphere, 14
Ionospheric reflection, 14
Isoadapter, 93-94

K

K$_u$ band, 35, 125

L

Launching satellites, 31
Leasing space on satellite, 34
Line of sight communication, 14, 15
LNA, 88
Long-distance communication, 15-17
Low-noise amplifier, 88-97

M

MATV, 125
Microwave, 126
Modulation, 12-13
Multielement yagi, 63, 66
Multiple microwave relay system, 37

N

NASA, 31
Noise
    temperature, 90, 126
        versus noise figure, 90
    versus frequency curves, 97
North American domestic satellites,
    32-33

O

Orbit selection, 21
Orthogonal polarizations, 62
Output signals, 118-119
Overall noise-figure
    performance, 96-97

P

Phase-locked loops, 111
Polarization efficiency, 48
Polar mount, 83-85

Power and bias circuitry, 96
Preemphasis of FM signal, 42
Prime-focus antenna, 66-69
Project Westford, 20
Propagation, 13-15

R

Radio
  frequency carrier, 12
  wave propagation, 16
Receivers, 105-107
Reflective elements, antenna, 63
Reflector platforms, 18
Remote controls, 114
RFI and EMI tight, 91

S

Satellite
  footprints, 22, 54
  locations, 134
  overview, 25
  positioning, 21-22
  transponder, 51-56
  TV broadcast stations, 40-46
  with solar panels deployed, 50
Single satellite coverage, 18
Solar panels, 50
Space loss, 47
Spherical antenna, 73-79
Spillover, 69
Spot beams, 21
Spreading loss, 47
Sputnik I, 20
SYNCOM I, 21

T

Telemetry, 49
Terrestrial microwave relay
  system, 36
Thermal noise, 94
Transponder, 35, 127
  bandwidth, 35
Tuner, 111-114
TV modulator, 119
TVRO, 28

U

Uncooled amplifier, 94
Uplink, 35, 36-48
  and downlink frequency plans, 43
  broadcast facility, 38
  makeup, 35
  propagation path, 46-48
  signals, 25

V

Vestigial sideband, 128
  filter, 38
  transmission, 38
VHF and UHF channel assignments, 41
VHF/UHF transmitter, 38-40
Video
  clamp, 44
  frame rate, 44
Voltage-controlled oscillator
  (VCO), 100

W

WARC, 25
Waveguide, 93, 128

Many thanks for your interest in this Sams Book about the world of Video. Here are a few more Sams Video Books we think you'll like:

You can usually find these Sams Books at better bookstores and electronic distributors nationwide.

If you can't find what you need, call Sams at 800-428-3696 toll-free or 317-298-5566, and charge it to your MasterCard or Visa account. Prices subject to change without notice. In Canada, contact Lenbrook Industries Ltd., Scarborough, Ontario.

For a free catalog of all Sams Books available, write P.O. Box 7092, Indianapolis, IN 46206.

THE HOME VIDEO HANDBOOK (3rd Edition)
Easily the nation's most popular, most respected book on the subject of home video recording! Shows you how to simply and successfully enjoy your home TV camera, videocassette recorder, videodisc system, large-screen TV projector, home satellite TV receiver, and all their accessories. Tells you how to hook everything up to make it do what you want it to do, how to buy the best equipment for your needs, how to make your equipment pay for itself, and more! By Charles Bensinger. 352 pages, 5½ x 8½, soft. ISBN 0-672-22052-0. © 1982.
Ask for No. 22052 . . . . **$12.95**

THE VIDEO GUIDE (3rd Edition)
If your work involves a hands-on understanding of video production hardware, this is your book. Tells you about standard and state-of-the-art videotape and VCR units for studio or portable use, industrial and broadcast cameras, videodiscs, editing systems, lenses, and accessories, and how they work. Then it shows you how and when to use each one, and how to set up, operate, maintain, trouble-shoot, and repair it. A classic reference guide for video professionals and ideal for those just learning, too. By Charles Bensinger. 264 pages, 8½ x 11, soft. ISBN 0-672-22051-2. © 1983.
Ask for No. 22051 . . . . **$18.95**

THE VIDEO PRODUCTION GUIDE
Helps you professionally handle all or any part of the video production process. Contains user-friendly, real-world coverage of pre-production planning, creativity and organization, people handling, single- and multicamera studio or on-location production, direction techniques, editing, special effects, and distribution of the finished production. Ideal for use by working or aspiring producer/directors, and by schools, broadcasters, CATV, and general industry. By Lon McQuillin, edited by Charles Bensinger. 352 pages, 8½ x 11, soft. ISBN 0-672-22053-9. © 1982.
Ask for No. 22053 . . . . **$28.95**

CABLE TELEVISION (2nd Edition)
Designed for the engineer or technician who wants to improve his knowledge of cable television. Helps you examine each component in a cable system separately and in relation to the system as a whole. Discusses component testing, troubleshooting, noise reduction, and system failure. Contains valuable information concerning fiber optics and communications satellites. By John Cunningham. 392 pages, 5½ x 8½, soft. ISBN 0-672-21755-4. © 1980.
Ask for No. 21755 . . . . **$13.95**

THE SATELLITE TV HANDBOOK
An easily read, amazing book that tells you what satellite TV is all about! Shows you how to legally and privately cut your cable TV costs in half, see TV shows that may be blacked out in your city, pick up live, unedited network TV shows that include the bloopers, start a mini-cable system in your apartment or condo complex, plug into video-supplied college courses, business news, children's networks, and much more! Also covers buying or building and aiming your own satellite antenna, and includes a guide to all programs available on the satellites, channel by channel. By Anthony T. Easton.
Ask for No. 22055

BASICS OF AUDIO AND VISUAL SYSTEMS DESIGN
Valuable, NAVA-sanctioned information for designers and installers of commercial, audience-oriented AV systems, and especially for newcomers to this field of AV. Gives you a full background in fundamental system design concepts and procedures, updated with current technology. Covers image format, screen size and performance, front vs. rear projection, projector output, audio, use of mirrors, and more. By Raymond Wadsworth. 128 pages, 8½ x 11, soft. ISBN 0-672-22038-5. © 1983.
Ask for No. 22038 . . . . **$15.95**

IF YOU'RE IN THE VIDEO BUSINESS...

And have a line of video products you need to sell, Sams can help you do it with your own AV/Video catalog.

It's neither difficult nor expensive to have a catalog produced with your name on the outside and your product lines on the inside, and it definitely makes you look good while it helps you sell.

Call 800-428-3696 or 317-298-5566 and ask the Sams Sales Manager for full details.